Project Administration for Design-Build Contracts

Other Titles of Interest

ALTERNATIVE PROJECT DELIVERY PROCUREMENT, AND CONTRACTING METHODS FOR HIGHWAYS
Edited by Keith R. Molenaar and Gerald Yakowenko
Provides an objective description of procurement methods that can improve the efficiency and effectiveness of public-sector project delivery.
ASCE Committee Report, 2007. ISBN 978-0-7844-0886-5

CONSTRUCTION CONTRACT CLAIMS, CHANGES, AND DISPUTE RESOLUTION, SECOND EDITION
Paul Levin
Merges the principles of construction law with practical advice to aid those involved in the construction claims process.
ASCE Press, 1998. ISBN 978-0-7844-0276-4

INTERPRETING CONSTRUCTION CONTRACTS: FUNDAMENTAL PRINCIPLES FOR CONTRACTORS, PROJECT MANAGERS, AND CONTRACT ADMINISTRATORS
H. Randolph Thomas and Ralph D. Ellis
Uses rules set forth by the judiciary and decades of practical experience to discuss troublesome contract clauses and present rules to construe them so as to avoid disputes.
ASCE Press, 2007. ISBN 978-0-7844-0921-3

PREPARING FOR DESIGN-BUILD PROJECTS: A PRIMER FOR OWNERS, ENGINEERS, AND CONTRACTORS
Douglas D. Gransberg, James E. Koch, and Keith R. Molenaar
Covers the basics of developing a design-build project, from developing requests for proposals, to preparing an effective proposal, to evaluating and selecting a source.
ASCE Press, 2006. ISBN 978-0-7844-0828-5

PREPARING REQUESTS FOR PROPOSALS AND SPECIFICATIONS FOR DESIGN-BUILD PROJECTS
ASCE Committee on Specifications, ASCE Construction Institute
Explains in detail the key steps in preparing a concise, well-written request for proposals.
ASCE Committee Report, 2008. ISBN 978-0-7844-0987-9

TEN COMMANDMENTS OF BETTER CONTRACTING
Francis T. Hartman
Outlines ten basic rules for improving contract performance.
ASCE Press, 2003 ISBN 978-0-7844-0653-3

THE ENGINEERS JOINT CONTRACT DOCUMENTS COMMITTEE (EJDCD) CONTRACT DOCUMENTS
Fair and objective standard documents that represent the latest and best thinking in contractual relations between all parties involved in engineering design and construction projects. For more information, please visit http://pubs.asce.org/contracts.

Project Administration for Design-Build Contracts

A Primer for Owners, Engineers, and Contractors

James E. Koch, Ph.D., P.E.
Douglas D. Gransberg, Ph.D., P.E.
Keith R. Molenaar, Ph.D.

ASCE PRESS

Library of Congress Cataloging-in-Publication Data

Koch, James Edward.
Project administration for design-build contracts : a primer for owners, engineers, and contractors / James E. Koch, Douglas D. Gransberg, Keith R. Molenaar.
p. cm.
Includes bibliographical references and index.
ISBN 978-0-7844-1075-7
1. Building—Superintendence. 2. Buildings—Specifications. 3. Construction contracts. I. Gransberg, Douglas D. II. Molenaar, Keith R. (Keith Robert III. Title.
TH438.K553 2010
690.068—dc22

2010023045

Published by American Society of Civil Engineers
1801 Alexander Bell Drive
Reston, Virginia 20191
www.pubs.asce.org

ISBN 978-0-7844-1075-2
Manufactured in the United States of America.

18 17 16 15 14 13 12 11 10 1 2 3 4 5

Contents

Foreword . . . . *ix*
by Peter Beck

Preface . . . . *xi*

ONE INTRODUCTION TO DESIGN-BUILD PROJECT ADMINISTRATION 1

Design-Build Background 1
Summary of Recent Design-Build Research. 4
Design-Build versus Design-Bid-Build: What Is Different 9
The Design-Build Project Team 20
Identifying the Preeminent Factor for Using Design-Build 27
Partnering the Design-Build Contract after Award. 29
Design-Build Success Factors 33
References 33

TWO SCHEDULING DESIGN-BUILD PROJECTS 37

Analyzing Contract Schedule Requirements 39
Design-Build Scheduling Conference 42
DB Schedule Imperatives 62
References 63

THREE DESIGN ADMINISTRATION IN DESIGN-BUILD 65

Changing the Design Administration Culture 66
Design Personnel. 71
Integrating Contract Design Input 75
Design-Build Charettes 81
Performance Criteria Interpretation 85
Design Submittals and Reviews 89
Drawing Numbering System 93
References 95

FOUR DESIGN QUALITY MANAGEMENT IN THE DESIGN-BUILD PROJECT 97
Design-Build Quality Assurance Model 98
Design Quality Defined 106
The Economics of Quality in the Design Phase 118
References 146

FIVE DESIGN ADMINISTRATION CASE STUDIES 149
Case 5-1: "When Is Design *Really* Design?" 149
Case 5-2: "Actions Speak Louder Than Words." 152
Case 5-3: "The Growing Design" 154
Case 5-4: "All the Right Moves" 157
Case 5-5: "Keep Looking until You Find the *Right* Answer." 159

SIX CONSTRUCTION QUALITY MANAGEMENT IN THE DESIGN-BUILD PROJECT 163
Construction Quality Defined 166
Owner's Role in Defining Construction Quality Requirements 168
The Economics of Quality in the Construction Phase 172
Construction Quality Management Planning 176
Verification and Acceptance Testing 183
Post-Award Construction Quality Plans 185
Construction Quality Personnel Requirements 186
References 188

SEVEN DESIGN-BUILD CHANGE ORDERS 191
Design-Build Change Order Entitlement 194
Design-Build Change Management 196
Design-Build Change Order Process during Design 200
References 217

EIGHT PROGRESS PAYMENTS DURING DESIGN AND CONSTRUCTION 219
Payment for Design Progress 221
Retainage and Final Payment 230
Limitation on Payments 235
Incentive Payments 240
Design-Build Project Cost Control 248
References 253

NINE CONSTRUCTION ADMINISTRATION CASE STUDIES 255
Case 9-1: "Trust Me. I Know This Will Work." 255
Case 9-2: "The Winning Inspector Is . . ." 257
Case 9-3: "If You Have Never Seen It, You Will Never Want It." 258
Case 9-4: "Were You Really Sure You Were Ready for This Project?" 261
Case 9-5: "Whom Are You Trying to Satisfy?" 264

Appendix A: Implementing Building Information Modeling in Design-Build Projects 267
by Tamera L. McCuen

Appendix B: Implementing Sustainability in Design-Build Projects 273
by Lee A. Fithian

Index 281

About the Authors 285

Foreword

I am both pleased and honored to be invited to write a foreword for a book on the topic of design-build contract administration. This type of book has been a long time in coming and provides much-needed guidance for both design-builders and—more importantly—our valued clients. I have had the pleasure of knowing Doug Gransberg for years through The Beck Group's involvement with the University of Oklahoma's Construction Science Division, where we have found many of our best employees. Doug's program has been one of a few that seeks to prepare entry-level designers and constructors with the skill set to be successful as integrated design-builders. We have also employed him as a design-build consultant to assist The Beck Group's efforts to better understand federal government opportunities. I have always found his in-depth and practical understanding of design-build to be unique, refreshing, and informative. I know Doug's co-authors, Keith Molenaar and Jim Koch, by reputation and through our mutual involvement in the Design-Build Institute of America. A book by this trio of experienced design-build researchers, practitioners, and consultants is written from a deep well of experience. Keith Molenaar was the first researcher in the nation to tackle this emerging area in the early 1990s, and Jim Koch worked with Doug in their previous careers in the U.S. Army Corps of Engineers, where they pioneered the Corps's design-build program. He has spent the subsequent years putting his experience to work for the nation's benefit as a director of the Rebuild Iraq program. Thus, the information contained in this book has literally been tested in the most challenging project delivery environments.

This is the opportune time for all practitioners to develop a keen understanding of the design-build delivery method. Up until now, too many design-build projects either appear to have been designed by the contractor or fail to embrace the unique opportunities to integrate the disciplines. We must learn to build upon our knowledge of the process on a project-by-project basis and ultimately consider marrying the disciplines within the same firm in order to amortize investments in technology, cross-training, cross-disciplinary processes, and so on over many projects. The

Integrated Enterprise will eventually leverage the benefits of the design-build process by eliminating the massive waste in traditional delivery methods and discover far better means and methods in the design and construction of projects.

Gransberg, Molenaar, and Koch have laid out a framework from which we can all learn as we proceed along this exciting journey. I urge you to take the time to understand what they are explaining and then seek to apply it in your own practice in a disciplined manner. Learn from your experiences, but open yourself to new practices and processes.

PETER BECK
MANAGING DIRECTOR, THE BECK GROUP
DALLAS, TEXAS
January 2010

Preface

The delivery of public and private construction projects using design-build (DB) continues to expand, encompassing virtually every sector of the construction market. Proven DB success in both the private and public sectors is fueling the growth of this project delivery method. As a result, many of the traditional relationships that exist in the architect/engineer/construction community are being reforged to permit them to operate within projects that are being delivered using alternative methods such as DB. Additionally, DB is no longer a new project delivery method; it is being institutionalized on a large scale throughout the world. As a result, many designers and builders will find themselves being drawn into DB projects due to owner pressure to compress project delivery time frames. Thus, these design professionals must be prepared to alter their business practices to accommodate the changed relationships within a DB contract, which the DB contract administration process to become a key attribute of project success.

Much of what has been written on the subject of DB is about the mechanics of the contracts. While this is necessary information, nothing has been published about how to actually administer the technical requirements of the DB contract. Additionally, consulting engineers, architects, and construction contractors are finding that new roles have been created with the advent of DB project delivery. The design community has felt the most effect of this business culture shift. With this project delivery system, the designer must decide whether to remain on the owner's team and assist it by preparing the DB request for proposal, or join forces with a construction contractor and become the designer-of-record (DOR) on the DB team. Construction professionals must also recognize and understand the impact of these new roles. The preconstruction period, when the designer and builder join forces to satisfy the owner's project requirements, creates an opportunity to implement both design creativity and innovative construction means, methods, and management. The DB approach creates a single point of responsibility for both design and construction and gives it to the DB team. Thus, this book will explore in depth the implications of the culture shift that is associated with the move to a scope-driven, performance-based delivery system.

In 1996 we developed a professional continuing education seminar entitled "Design-Build Contracting" under ASCE's auspices, and since then have delivered it across the nation and in several foreign countries. The one item that seemed to be common to all those classes was the pressing need by engineers who worked for owners, for consulting firms, and for construction companies to better understand the technical side of the DB process and to be able to put it into context with the contractual side, with emphasis on DB project execution. Based on this experience, this book was developed to complement the book we published in 2006, *Preparing for Design-Build Projects: A Primer for Owners, Engineers, and Contractors*, which focused on the pre-award activities of owners, designers, and builders in advertising, proposing, and awarding a DB project. This book is restricted to discussing the post-award activities involved in actual design and construction contract administration.

The first chapter sets the stage by furnishing an introduction to the topic of DB contract administration and a synopsis of the relevant literature on the subject. Because DB project delivery is often selected to achieve an accelerated schedule, the second chapter discusses techniques and practices for developing a cogent schedule of all design, logistical, administrative, and construction activities. The next two chapters focus on design administration and design quality management in the project. This is followed by a synthesis of the salient points of the first half of the book illustrated through a series of actual DB case studies. Next, the book moves on to the construction phase with chapters on construction quality management, change order procedures, and progress payments in the DB context. This is followed by another series of case studies that relate to interaction between the owner's Request for Proposal (RFP) and the design-builder's proposal. Finally, the book concludes with two essays contributed by experienced design-builders on building information modeling in DB and implementing sustainability.

This book is intended to be a resource for owners, engineers, construction contractors, and architects who find themselves in need of guidance in administering a DB project. It is unique in that previous books have either taken a global approach to the subject or have concentrated on the legal aspects of the contracts themselves. This book is intended to help those professionals who must actually do the designing, the building, and the contract administration. It tries to lay out all the options in a manner that comparatively highlights the advantages and disadvantages of each option so that the reader has all the information necessary to make the business decisions inherent to the DB process.

We would like to thank all those other DB professionals around the country who let us bounce ideas off them and helped us gather case studies. Special thanks go to Professors Tammy McCuen and Lee Fithian of the University of Oklahoma for their insightful essays regarding the use of building information modeling (BIM) and integrating sustainability in DB projects.

ONE

Introduction to Design-Build Project Administration

In 2006 we published a book entitled *Preparing for Design-Build Projects: A Primer for Owners, Engineers, and Contractors*, which focused on the activities of owners, designers, and builders in advertising, proposing, and awarding a design-build (DB) project (Gransberg et al. 2006). That book's purpose was to relate the DB project delivery method body of knowledge, from concept to procurement. The goal of that book was to help set up owners and DB teams for success. This book will pick up that thread to help ensure successful project delivery through the design, construction, and commissioning phases of the process. As its title suggests, this book will focus on DB project administration post-award. While these two books are intended to be companions and to function as a set that covers the life cycle of typical DB projects, we recognize that some readers may choose to read only this one. As a result, a short introduction to DB is in order to set the stage and effect the transition from the first book to this one.

Design-Build Background

The move from traditional design-bid-build (DBB) project delivery to nontraditional forms of integrated project delivery has been a controversial one in the United States. "Integrated project delivery" is defined as a project delivery method where the project's builder is involved in the design process. While there are many terms-of-art that are used in the literature to describe various forms of integrated project delivery, the two most common forms are "construction manager-at-risk" and "design-build." In both project delivery methods, the owner selects the builder at a point in time before the design is complete and seeks to involve that builder in the design process through the use of formal design reviews that seek to enhance constructability, expedite the schedule, and maintain budget. While the use of integrated project delivery remains controversial with some owners, particularly in the public sector, its acceptance is becoming more general across the nation. This book will confine its discussion to DB project delivery and, specifically, the actions that take place after the award of a DB contract to complete both the design and construction in an integrated fashion.

The Design-Build Institute of America (DBIA) believes that the majority of nonresidential construction projects will be delivered using DB in 2010 (Eby 2005). Additionally, a survey in 2004 found that construction companies expected 50% of their revenues to come from DB projects in 2006 (Zweig-White Research 2004). That survey also found that 80% of all design and construction firms surveyed in the United States expected an increase in their business derived from DB projects over the next five years. According to DBIA, the percentage of nonresidential construction projects being delivered by DB has been increasing steadily for the past two decades, from an estimated $18 billion in 1986 to more than $250 billion (DBIA 2005). This change in project delivery methods began in the 1960s in the private sector on commercial construction projects where the financial benefit of reducing the project delivery schedule outweighed the risk of beginning construction without a complete design (Gransberg et al. 2006; Loulakis 2003; Molenaar et al. 1999). Over the next 20 years, the DB approach was adopted by the public sector as a method for delivering revenue-producing projects, such as toll roads and bridges, as well as an effective means to expedite the procurement of emergency reconstruction after natural disasters such as the interstate bridges demolished by hurricanes in Florida. In 1996, the Federal Acquisition Reform Act was passed and specifically provided both regulation of the use of DB on federal projects and authority to utilize the DB delivery method without seeking special permission (Gransberg et al. 2006).

Since 1993, the DBIA has closely followed public procurement laws regarding DB. It has documented the trend of expanded statutory authority to public sector engineering and construction agencies to legally use DB in all types of construction procurements. The architectural/building sector has led the infrastructure sector in terms of implementing the use of DB. For instance, states including California and Oklahoma have authorized its use on public buildings without extending broad DB authority to their Departments of Transportation (DOTs). Nonetheless, in the past 10 years, DB transportation projects have been constructed in more than 35 states, as stated in a Federal Highway Administration (FHWA) report to the U.S. Congress on the effectiveness of DB delivery in federally funded projects (FHWA 2006). DB has been used successfully on toll roads and mass transit projects where the revenue generation potential forms a politically acceptable argument for achieving an early opening by compressing the traditional delivery period to as short as possible. Thus, the use of DB project delivery in the transportation sector is growing across the country as well.

By 2004, the FHWA had authorized more than 300 DB transportation projects in 32 states, worth nearly $14 billion (FHWA 2006). By 2002, the Florida Department of Transportation (FDOT) alone had awarded 49 DB projects for nearly $500 million worth of work, and estimated that DB cuts 30% off the traditional project delivery period (Peters 2003). Another study reported that nearly $3 billion worth of water/wastewater projects were being delivered annually using DB

(DBIA 2000a). Adding the uncounted number of public building, utility, and other infrastructure DB projects completed by county and municipal public agencies, as well as the public–private partnerships (PPPs) that deliver critical infrastructure such as toll roads, toll bridges, and water/wastewater projects, the nationwide market for DB project delivery is enormous. Such meteoric growth in roughly 20 years solidly confirms that DB project delivery supplies real benefits to the public and private owners that implement it. The FHWA recently completed a seminal report on DB project delivery in transportation that eloquently articulates this:

> The greatest motivation and realized benefit to a contracting agency of using design-build . . . is the ability to reduce the overall duration of the project development process by eliminating a second procurement process for the construction contract, reducing the potential for design errors and omissions, and allowing for more concurrent processing of design and construction activities. . . . (FHWA 2006)

As previously mentioned, the building sector has led the growth of DB project delivery. The building sector encompasses a wide range of project types falling under the general heading of "vertical" construction. Project types may include categories such as high- and low-rise public offices, courthouses, correctional facilities, schools and related educational facilities, warehouses and distribution centers, public and military housing, parking structures, stadiums and sports facilities, libraries, and museums. Owners or users of these facilities range from federal, state, and local governments, to public authorities and agencies, school and university boards, municipalities, and county commissions. Owners have selected DB delivery for all of these types of projects. The DBIA surveyed federal building owners regarding their motivations for using DB; all responded that schedule compression was a primary criterion, and most cited the criteria of cost, quality, and user management as well (DBIA 2000b). The DBIA's findings mirrored a University of Colorado/National Science Foundation study showing that schedule reduction was the overwhelming reason for the selection of DB on building projects in both the public and private sectors (Songer and Molenaar 1996).

While the "horizontal" sector has lagged in DB growth, it has similar motivations for using the method. Completing needed construction projects as fast as possible is the paramount factor in transportation infrastructure projects, and nowhere is it more critical than on congested urban freeway projects. User costs of congestion due to delays, increased toxic emissions, and other factors on urban freeways can exceed $10,000 per lane-mile/day (Hicks and Epps, 2000). Thus, a typical eight-lane freeway in a major city like Los Angeles or New York would incur a user cost of $80,000 per mile per day, and a hypothetical five-mile urban freeway upgrade project would incur user costs on the order of $400,000/day. Taking the analogy a step further, and in keeping with the above FHWA statement, one could calculate that the 30-day period normally used to bid a traditional DBB project would cost the commuters $12 million on that five-mile stretch of freeway. DB would generate a significant benefit by delivering the needed upgrade a full month earlier than it would have been delivered using traditional DBB.

Summary of Recent Design-Build Research

Our previous book covered much of the significant research in DB project delivery. Thus, an update of the research that has been completed since that book was written will be included here to bring the reader up to date in this important area. The major change that has been seen in DB research is its move from the building sector to the transportation sector as public laws have been enacted across the nation allowing its use in major engineered projects. The focus of the research has become more specific, looking at such subjects as the impact of DB on the owner's workforce and how the quality management process must be altered to accommodate DB project delivery. Additionally, research projects in DB have been funded by professional societies such as the DBIA and the Society for Marketing Professional Services to transfer knowledge from the owner's perspective in a manner that is valuable to design-builders. The following short review will be limited to those projects that speak to the DB project administration process and thus furnish a theoretical foundation for the remainder of this book. A comprehensive summary of DB research prior to 2003 can also be found in the book *Design-Build for the Public Sector,* Chapter 3, "Examining the Performance of Design-Build in the Public Sector" (Loulakis 2003).

The most significant study in the past few years was a report to the U.S. Congress commissioned by the FHWA on the effectiveness of DB in transportation projects. It is arguably the most comprehensive study ever completed on the subject. It looked at nearly every DB project that had been authorized by the FHWA and compared those projects to traditional projects using most conventional metrics of project performance. It addressed project performance in the areas of cost, schedule, and quality, and can be elegantly summarized by the following excerpt from the report:

> On average, the managers of design-build projects surveyed in the study estimated that design-build project delivery reduced the overall duration of their projects by 14 percent, reduced the total cost of the projects by 3 percent, and maintained the same level of quality as compared to design-bid-build project delivery. (FHWA 2006)

In this single statement, this study makes a strong case that the benefits of compressed schedule and marginal cost savings were achieved in more than 300 projects across the nation without sacrificing quality of the constructed product.

The emergence of DB contracting on the national transportation scene has certainly been controversial. The emotions associated with the paradigm shift required to implement it have run high. However, the results of the FHWA study should serve to temper those feelings by addressing the concerns of public and private owners with facts.

A 2003 study for the National Cooperative Highway Research Program (NCHRP) of the Transportation Research Board of the National Academies recorded the perceptions of state DOT upper managers regarding the benefits experienced from implementing DB:

> With respect to design-build projects, the conventional wisdom among those interviewed was that the dollar cost to the agency appeared to be about the same as conventional approaches, but that design-build projects opened to the public much sooner. Time, of course, is money and therefore this suggests that indeed there is a considerable saving to the public in the form of user cost benefits due to significant schedule accelerations resulting from this outsourcing method. (SAIC 2003)

That same NCHRP study goes on to cite the Maryland State Highway Administration's experience with its first 10 DB projects where it "reduced the average time for design and construction by approximately one year, as compared to the traditional design-bid-build method. The program has also consistently produced final products with less than 1% in change orders—significantly less than the [DBB] program as a whole" (SAIC 2003). Thus, DB not only appears to improve project delivery time but also enhances *project cost certainty*. This tracks with research conducted 14 years ago that found that the top two reasons for public owners selecting DB project delivery were to compress the project schedule and to establish the project cost at an earlier point in time (Songer and Molenaar 1996).

Project cost certainty and early cost establishment are very important to public infrastructure projects because they significantly affect the number of capital projects an agency can award in a given year within its established budget. For instance, an early study of FDOT's first series of DB projects found that those projects were awarded at a 4.5% premium above low bid but only experienced 2% cost growth. Conversely, FDOT DBB projects experienced 8.5% cost growth from the low-bid award price (Ellis et al. 1991). Thus, there was a marginal 2% cost savings on the DB projects. More important, however, was the effect on FDOT's ability to manage its capital improvements budget. Under the traditional system, FDOT would have had to set aside 8.5% of each project's authorized budget to pay for cost growth due to quantity overruns and change orders. Its DB experience would require it to set aside only 2%. Thus, it could award a far greater percentage of its authorized budget, which presumably would result in a larger number of total infrastructure projects being built in a given fiscal year and a more efficient use of available capital (Gransberg and Villarreal 2002).

Actually, a number of factors have driven the increased use of DB to deliver much-needed infrastructure projects. Hancher and Werkmeister (2001) cite the growth in U.S. population creating increased travel demand as the root cause for an increased requirement for construction of new infrastructure and reconstruction of aging infrastructure. They further argue that the traveling public's demand for "better and quicker service," combined with an overall reduction in

DOT workforces in the era of "reinventing government," has pushed public transportation agencies to explore alternatives, and DB is one way that can be accomplished.

The detractors of the DB project delivery method for highway projects who voiced their opinions in the early 1990s have been quieted by the overwhelming momentum of DB use across the country. The FHWA provided a lengthy and rigorous comment period on its Design-Build Final Rule (FHWA 2002). The FHWA considered these comments and concerns and then published rules that allow, but do not require, all recipients in the Federal-Aid Highway Program to use the DB contracting method just as they would the traditional DBB contracting method. The majority of states are now using DB (FHWA 2006). A number of states have developed large and successful DB programs, while others are still growing their programs or applying DB on individual projects that are well suited for this delivery method. The benefits of faster delivery and greater cost certainty have been repeatedly validated, while the concerns of lowered quality are inconclusive at best.

Quality management in DB is the heart of the issue, and concerns about how this vital piece of the design and construction process will be accomplished have been voiced since the inception of DB. Quality of the finished product is a significant measure of success in any construction project. In its previously cited report to Congress, the FHWA addressed this issue in its survey of highway agencies and summarized its findings as follows:

> Design-build does not appear to be a threat to the quality of highway projects. Indeed project contracting agencies expressed equal satisfaction with the results of design-build and design-bid-build projects, suggesting that the choice of project delivery approach is neither a determinant of nor a threat to project quality. Overall contracting agency satisfaction was highest when design-build was used for large projects, when lower levels of preliminary design were performed prior to the design-build contract, and when contract selection was based on best value. (FHWA 2006)

Although owners have been satisfied with the end-product quality, there are many challenges to implementing DB quality systems in an historically DBB environment. In 2006 NCHRP funded a study of quality assurance in DB transportation projects that was completed by two of the authors of this book (Gransberg et al. 2007). It looked at results from a general survey on DB quality management that produced 76 responses from 47 states. That study also analyzed the contents of DB solicitation documents from projects with a total contract value of more than $11.5 billion from 26 transportation agencies across the country. For further verification, that study also analyzed the contents of DB policy documents from 17 states, and the data were collected from a brief survey on DB quality perceptions from those states. It found that it was necessary to add a new level of quality management activity to the traditional quality assurance system to cover an owner's oversight responsibilities for delivering a quality project when it chose to delegate all the quality assurance and quality control responsibilities to the design-builder. That study coined the term "project quality assurance" (PQA) to label this DB-specific quality management responsibility. The study also found that owners were not paying sufficient attention to design quality management because the public owners in that study were accustomed to doing their own design work on DBB projects. Thus, the study found that there were new roles for both owners and design-builders during the project administration process that affected the workforce involved in DB projects.

Another concern of DB detractors regarding public infrastructure projects relates to its perceived impact on the public workforce. The Keston Institute for Public Infrastructure Policy and Finance commissioned a study by two authors of this book on the impact that implementing DB has on the workforce (Gransberg and Molenaar 2007). That study found that public agencies that implement DB do not stop using DBB project delivery. In fact, the average public transportation construction program contained only about 10% of the projects being delivered using DB. Additionally, the study found that owners were not reducing their workforce by using DB to outsource design and construction engineering work. In fact, it found that to successfully use DB, an owner had to have a more competent and experienced workforce that was able to expeditiously make decisions based on professional judgment. The study also found that the roles of the owner and design and construction personnel changed on DB projects. The in-house professional engineering workforce shifted from doing detailed design to design review, and the construction workforce increased its focus on quality assurance.

One of the major challenges facing DOTs and design-builders in implementing DB is the change that must take place in the culture of both parties. This is described well in an evaluation report given by Postma et al. on the I-15 DB project in Utah:

> The Owner felt that one of the biggest challenges to the QC and QA program was "breaking the mold" of the traditional roles of the Contractor and Owner. The Owner's personnel had all come from the "catch and punish" culture. Likewise the Contractor's personnel came from a similar background. To change philosophies to a more proactive quality role by the Contractor and a less controlling oversight role of the Owner was a significant challenge. Most personnel assigned to the project by either party had worked under traditional systems for many years, and this was the first experience with this type of project. (Postma et al. 2002)

As both owners and design-builders become more familiar with DB, culture change will become less of an issue.

Another issue that has confronted DB projects since their implementation has been the concern that constructors, who are the leads in many DB projects, would pressure the designers, who are often subcontractors, into sacrificing quality for higher profits. According to the FHWA report to Congress already referenced

(FHWA 2006), this has not been the case. However, having a well-defined quality management system is one way to address some of these fears. This is even more important on high-profile projects delivered using DB. A quality management system adds credibility and assurance for all involved, from the constructors, to the DOT, to the public users (Pantazides 2005). Thus, the research supports the use of DB project delivery on projects where the owner will accrue tangible benefits from its use.

Given the spread of DB across the entire industry, it is important that owners, designers, and builders understand the process and its impact upon their stake in the design and construction process. In our 2006 book we discussed these aspects in the pre-award process that owners and design-builders use to arrive at a DB contract. The remainder of this book will concentrate on identifying the stakes held by each player in the DB project administration process.

It is important to note that DB project delivery, while seeming to be linearly divided into procurement, design, and construction phases, is really an amalgamation of all three activities that occur in parallel rather than in series. DB intentionally addresses procurement, design, and construction concurrently. For instance, the owner actually starts the design phase when it develops the DB Request for Proposal (RFP), making many design decisions before requesting proposals. The competing design-builders amplify the owner's design during the procurement phase through their technical proposals. If the contract is awarded on a negotiated basis, both design and construction activities can be touched on during those pre-contract negotiations. Once the contract is awarded, both design and construction can commence as soon as the owner issues its Notice to Proceed. Therefore, approaching the subject of DB project delivery requires one to surrender many time-honored beliefs that really only apply to DBB project delivery, which is by design an inherently linear process. In fact, the DB process is occasionally referred to as "design and build." Nothing could be further from the truth. In DB project delivery there is no clear demarcation between the end of design and the start of construction unless the owner refuses to allow the design-builder to proceed with construction before the design is finalized. And even in that case, the design-builder will begin many traditional construction activities, such as locking-in materials contracts at target prices and making arrangements for equipment and subcontractors for the project, before the design is concluded and approved.

Therefore, one should read this book with the idea that DB project delivery is a very dynamic environment in which all parties to the DB contract are striving to achieve the same goals by working together, rather than segregating each other's efforts. As indicated in the above FHWA quote and research by other authors (Songer and Molenaar 1996), the major reason why project owners decide to use DB project delivery is to compress the project delivery period. This compression is accomplished by concurrently performing tasks that have traditionally been nonconcurrent. Thus, DB project administration must be about developing a system where the project delivery schedule is facilitated while achieving the budget goals as well as high project quality.

Design-Build versus Design-Bid-Build: What Is Different

While users of building or infrastructure projects will likely never know which delivery method was used to complete the project they use, there are some fundamental differences for the project teams that deliver them. The primary differences involve the contracts, the use of performance criteria, the design review philosophy, DB project administration, constructability, DB project control, document control, safety oversight, and the high level of trust in DB project administration.

Differences in the Contract

The fundamental difference between DB and DBB is in the contract, specifically in the way risk and responsibility for design details are shifted from the owner to the design-builder. In a DBB contract, the owner hires a designer to develop the final construction drawings, as depicted in Fig. 1-1. The owner essentially "owns" the details of the design and as a result furnishes an implicit warranty that the plans are constructable and free from design errors and omissions. In a DB contract, however, the design-builder assumes "ownership" of the details of the design and is in turn responsible for providing both design documents and a constructed

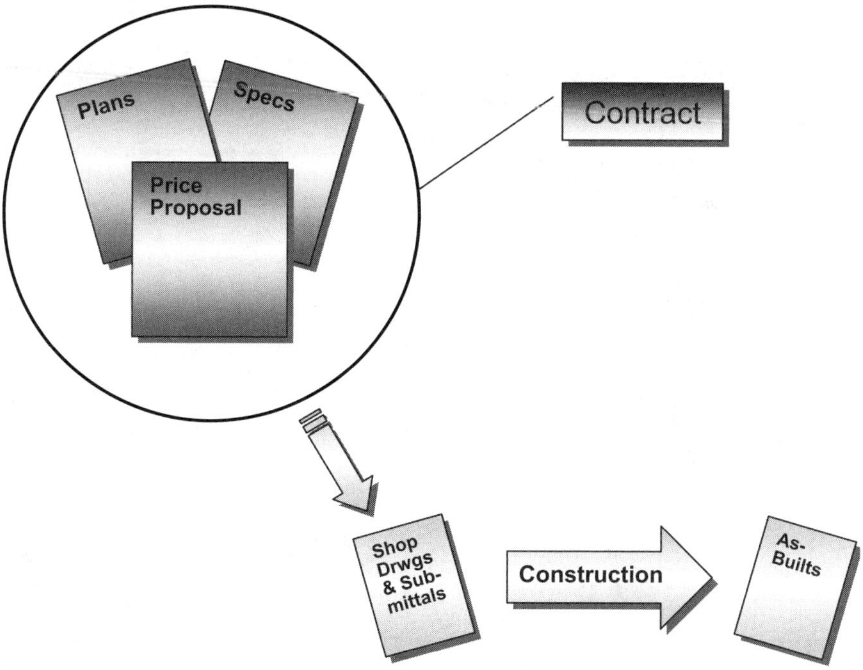

Figure 1-1 DBB contract model.

facility that are free of defects. In both delivery systems, the designer-of-record (DOR) is legally obligated to deliver a project that meets all applicable codes and standards within a reasonable standard of care. Figures 1-1 and 1-2 show simple models of competitively procured DBB and DB contracts and their places in the design and construction process. Admittedly, the models are an oversimplification of the process, but they very clearly convey several important nuances of the two delivery systems.

The underlying difference in the contract models is how and when the construction plans and specifications are delivered through the contact. Contracts are used to distribute risks between the parties in the project. Table 1-1 was developed by the Washington State Department of Transportation (WSDOT) and shows how it sees the change in risk distribution between DBB and DB projects. One can see in the table that in the traditional system, the final plans and specifications (often called the "construction documents") form the technical basis of the contract and define the detailed scope of work, along with the price proposal. A fundamental assumption of this system is that the plans are complete, constructable, and free from defects—often a difficult task. In DB, the plans and specification are a deliverable under the contract; the owner's RFP and the corresponding design-builder's technical and price proposals form the technical basis of the contract. This is a fundamental paradigm shift.

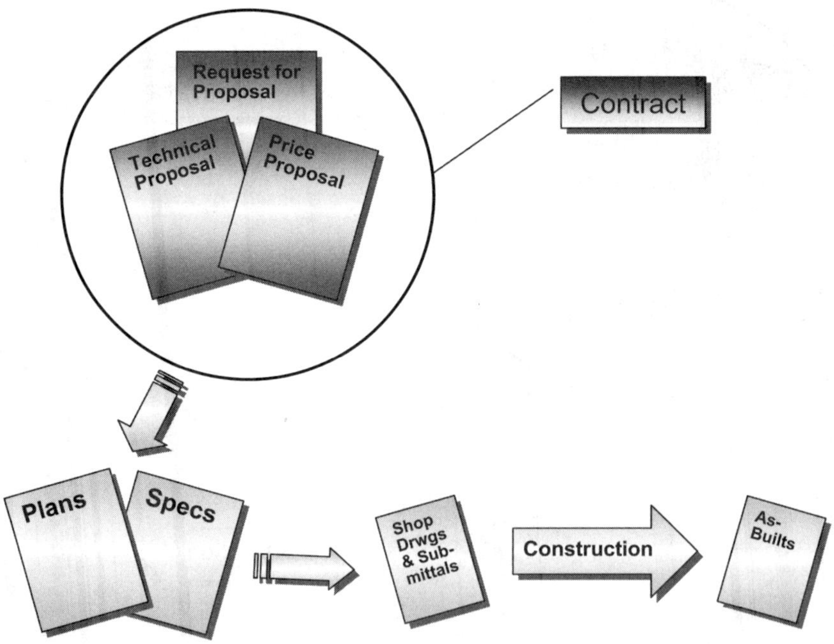

Figure 1-2 DB contract model.

Table 1-1 Washington State DOT Design Build Project Risk Matrix

	Design-Bid-Build			Design-Build Process	
Risk	**Owner**	**Shared**	**Contractor**	**Owner**	**Design-Builder**
Design Issues					
Plan conformance with regulations/guidelines/RFP	X				X
Plan accuracy	X				X
Design criteria	X			X	
Conformance to design criteria	X				X
Design review process	X				X
Design QC	X				X
Design QA	X				X
Owner review time	X			X	
Changes in scope	X			X	
Constructability of design	X				X
Construction					
DBE compliance			X		X
Safety/Safety QA			X		X
Construction quality/Workmanship			X		X
Schedule			X		X
Materials quality			X		X
Materials documentation			X		X
Material availability			X		X
Initial performance requirements of QA plan	X			X	
Final construction/materials QC/QA plan	X				X
Construction/materials QA	X				X
Construction QC			X		X
Construction QA procedural compliance auditing	X			X	
Construction IA testing/inspection	X			X	
Construction staking		X			X
Erosion control		X			X
Spill prevention		X			X

Table 1-1 (*Continued*)

	Design-Bid-Build			Design-Build Process	
Risk	**Owner**	**Shared**	**Contractor**	**Owner**	**Design-Builder**
Shop drawings			X		X
Equipment failure/breakdown			X		X
Work methods			X		X
Early construction/At-risk construction		X			X
Community relations	X			X	
Performance of defined mitigation measures	X				X
Warranty	X				X

Source: Adapted from WSDOT (2004).

One oversimplification in Figs. 1-1 and 1-2 concerns how the DB contract is procured (how the design-builder is hired and prices the work). The owner's RFP can range from a verbal request to one single design-builder, to a formal competition between design-builders from a published RFP that may contain boilerplate contractual language and substantially complete plans and specifications. Likewise, the technical and price proposals can vary greatly in detail, depending upon the owner, the design-builder, and the project. Pricing arrangements at the time the design-builder is hired can range from cost-reimbursable contracts to lump-sum price proposals that resemble a bid in the DBB process. Thus, when executing a DB project, all parties must constantly refer back to the documents produced during the procurement phase for guidance on how the physical design and construction were defined for contractual purposes. Again, the reader will find several "chicken or egg" situations throughout this book. It is impossible to divorce the actions taken or decisions made during the procurement phase from the post-award design and construction phases. This is particularly true concerning the issue of interpreting performance criteria used in the RFP during both design and construction.

Performance Criteria

In DBB, requirements are communicated to the constructor through complete drawings and specifications. In contrast, the nexus of communicating the requirements of a DB project is the owner's development of definitive performance criteria in the procurement phase, and their promulgation during design and construction.

These criteria serve to articulate the quality, cost, and schedule requirements for a given project and become the foundation for the DB contract. The contract model in Fig. 1-2 describes a salient aspect of DB project delivery: the concept that definitive, project-based performance criteria rather than comprehensive construction plans and specifications describe the project scope. This is another paradigm shift from DBB for owners, designers, and builders.

Owners must remember to define the project scope in terms of performance criteria rather than detailed drawings. It is often difficult to alter an owner's culture where its contract documents and administrative procedures may have been developed around the concept that complete construction drawings would be present at the time of construction procurement. In the public sector, these procurement cultures are long-lived; many were developed over the last five decades and they will not be changed easily or quickly. Architects, engineers, and constructors also need to learn how to work from a scope defined by performance criteria rather than complete construction documents. Designers are forced to develop a new sense of discipline to be able to design to both a budget and schedule defined in the RFP or the technical or price proposal. Additionally, they must continually monitor the design process to ensure that the quantities of work they generated during proposal preparation are not exceeded during design development. Constructors need to be attentive to the owner's needs and changes in scope, from which they were insulated in a DBB environment. All parties must learn how to embrace these performance criteria as the definitive project scope, or else risk costly scope creep as the project proceeds to completion.

Our previous book defined performance criteria as follows:

> A rule by which the effectiveness of operation or function is judged and its value measured. (Gransberg et al. 2006)

So, keeping in mind that the operative words in the definition are "effectiveness of operation or function," "judged," and "measure of value," the parties to a DB contract can move forward and interpret the performance criteria used to award the project on a best-value basis to the winning design-builder. These criteria set out the rules by which the evaluation panel and, post-award, the owner's project personnel will judge the effectiveness of the design-builder's design and construction. These criteria furnish the standard against which the value of each proposed design solution for every feature of work in the project will be measured. The net effect is to prescribe the standards for each of the project's various components. It can truly be said that having a clear understanding of the cogent performance criteria sets the stage for a successful project.

Design Review Philosophy

The next major difference between DBB and DB that occurs during project execution deals with the philosophy used by the owner to conduct design reviews. Because the design-builder now owns the details of design, the owner's design submittal reviewers are no longer conducting a technical review of the design. Their role is changed to the level of a compliance review. The technical review with its attendant checking of codes and calculations should occur as a part of the design quality control process implemented by the design-builder's DOR; see Chapter 4 for detailed discussion of this. The major reason for this shift is the need to ensure that the legal design liability remains squarely on the shoulders of the design-build team. An owner that does not train its reviewers on the difference between a "technical review" and a "compliance review" risks unintentionally assuming design liability as the DOR implements specific review comment directives. Thus, the issue circles back to the performance criteria established during the RFP process and the design reviewers' understanding that their role is to compare the design-builder's design solutions with those criteria and decide whether they comply with the published criteria.

The design reviewers must understand that the DOR must design to the budget and schedule that are specified in the DB contract. Additionally, it is incumbent on the design-builder to ensure that no design submittal is given to the owner that has not been checked to ensure that its contents meet both budget and schedule constraints. This is necessary to avoid the unnecessary appearance of "bait and switch" when a change is required for a previously approved design detail, due to either cost or availability issues. Mutual trust is essential to DB project success, and the owner's greatest concern is that the integrity of the design process will be compromised by financial considerations post-award. Thus, it is important that, when the design-builder shows its owner a product of the design process, it be able to assure that owner that the project will be constructed in exactly this manner if the design is approved. The issue of trust in DB project administration will be addressed in greater detail in the Trust in Design-Build Project Administration section later in this chapter.

Design-Build Project Administration

Although DB project administration is the overall focus of this book, it is appropriate to first discuss the high-level differences in this area from the traditional DBB design and construction administration processes before getting into the details.

The first point of difference is the fact that DBB project administration processes were developed with the idea that only design activities would occur during the design phase and only construction activities would occur in the construction phase. This resulted in a linear model for project administration wherein only designers and their product would be administered, followed by the constructors and their product. To successfully administer a DB project, however, this linear thinking on the part of the owner as well as the members of the design-build team must be laid to rest. Figure 1-3 shows the conceptual time lines for DBB and DB project administration; one can easily see that if an owner has separate design and construction administration systems, they will both need to be up and operating throughout the delivery of a DB project.

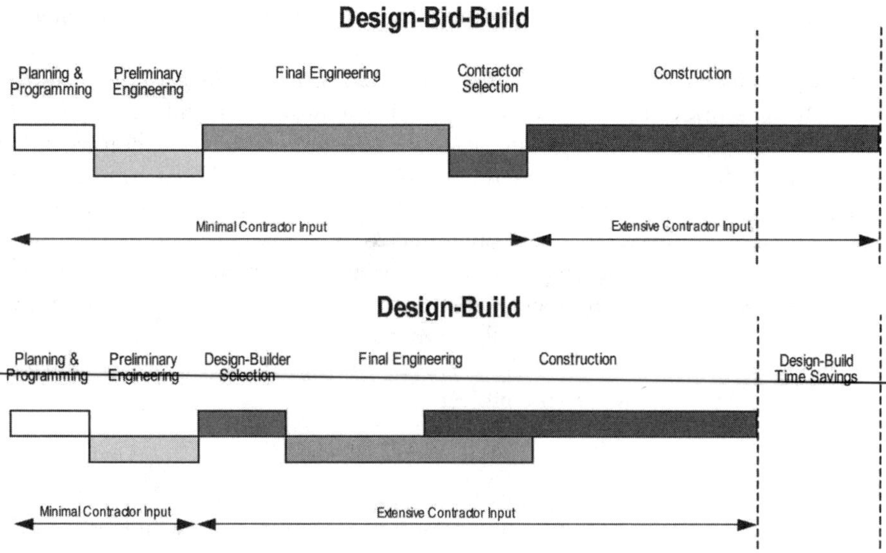

Figure 1-3 Conceptual timelines for DBB and DB project administration.

Constructability

One of the oft-cited benefits of DB project delivery is improved constructability based on extensive constructor input during the project's design (FHWA 2006). Figure 1-3 shows that this input actually begins during the procurement phase as the constructor works together with the DOR to develop the DB technical proposal that forms the technical core of the DB contract. This input continues throughout the design process and into construction. Thus, unlike in DBB, a DB project's design can be thought to be under continuous constructability review. In DBB, if there is a constructability review, it typically takes place at a single point in time after the design is complete. If that review uncovers opportunities to improve the project's constructability, a redesign effort will be necessary, which will result in the loss of previously expended design effort. Thus, in DB, the constructability process is perceived to add value to the final project, whereas in DBB, constructability reviews are not usually welcomed by the project's designers because they are perceived to be criticism rather than assistance.

To accrue the benefits of continuous constructability review, the project design administration system must be established in a manner that facilitates this critical process rather than discourages it, as is commonly done in DBB. One method for doing this is to adopt a contractual concept of "design commitment"; this is discussed in detail in our previous book as well as in Chapters 3, 7, and 8 in this book. Design commitment is essentially made when a working drawing and/or specification becomes a construction document approved by the DOR for construction.

In other the words, the DOR is indicating that no further design work is contemplated and that this feature of work is ready for final owner review. Additionally, the design-builder is indicating that the committed design product has been priced and that the work can proceed according to the approved schedule. Thus, in DB the owner's representative has a yardstick with which to measure the construction progress and authorize progress payments. Any changes after design commitment require the owner's approval. Before design commitment, either the DOR or the design-builder can make changes as required to meet budget and schedule constraints as long as they comply with the technical requirements contained in the contract. Thus, design commitment essentially acts as a payment mechanism in the design administration system, as well as a means to communicate the progress of design effort in a DB project. It also serves to enhance the sense of mutual trust and honesty that is essential to the project.

Project Control

The DB project administration system must support control of the budget, schedule, quality, safety, and overall documentation. We have found that the key to DB project control is recognition that the budget and the schedule are fixed before the ultimate details of project quality are determined. Our previous book described this as the "three-legged stool" shown in Fig. 1-4. Thus, to appropriately control the project from both the owner's and design-builder's perspective, all the project control systems must be developed with the idea that the variable leg in the stool is the details of the design. Once those details change from pro-

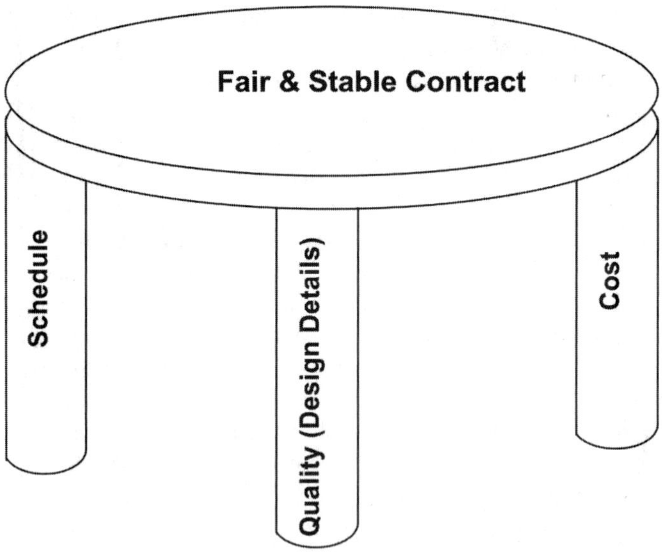

Figure 1-4 Project delivery concept.

posed to approved (i.e., from the proposal design to the final design), the remainder of the project control system will function similarly to traditional systems. However, as each of the five project control areas is established, the entities involved must check them to ensure that if one control system, such as budget, indicates a problem, it is possible to address that issue without creating another issue in the design or construction quality control system. This book devotes specific chapters to DB project control for the schedule and budget, as well as two chapters on quality. Therefore, a brief description of the document and safety control systems is in order here.

Document Control

Document control in DB includes both design and construction documents as well as the universe of contract communications that occur during any type of construction project. The major difference between DB and DBB in this area deals with the need to be able to track the approval of design details directly into the constructed product. In a fast-track DB project, this becomes even more critical. The review and approval of design submittals by the owner will always be on the critical path, and the timely disposition of these documents should be the primary thrust of the DB document control system. Additionally, as will be discussed in a subsequent chapter, determining the appropriate number of design reviews based on project schedule constraints has become prevalent throughout the industry. A recent study of public owners in the transportation industry found that 15% did not require intermediate design reviews before the final construction documents were reviewed, and another 56% mandated only one intermediate review before the final review (Gransberg et al. 2007).

Concerning the architectural industry, it is interesting to note that the U.S. Army Corps of Engineers (USACE) recently changed its policy for DB design reviews, reducing the number of reviews from four (30%, 60%, 90%, and final) to two (intermediate and final) (USACE 2006). The reason for the change was to reduce the potential for delays due to waiting for government reviews. In a personal communication with the authors, Joel Hoffman of USACE explained the rationale as: "[The] philosophy is that once the designer of record approves construction and extension of design submittals, the builder can proceed—don't wait on us, unless there is a specific government approval required." Thus, one critical issue regarding determination of the appropriate number of design reviews is the need for the design-builder to maintain an aggressive schedule. If the project is not schedule-constrained, the owner can afford to inject more design review points; however, design reviews should be minimized on a fast-track project.

A similar philosophy needs to be established for the control of construction submittals. In DBB, both the design-builder and the owner strive to complete as many of the typical construction submittals as part of the design process as possible. This involves crafting subcontracts that require early submission of subcontractor details of design, such as fabrication drawings and material submittals. For those that cannot be accomplished until after the final design is released for construction, a system to expedite the assembly review and approval by the appropriate authority is essential. Conversely, in DB the DOR rather than the owner will normally be assigned a large part of the construction quality assurance responsibility to review construction submittals and to verify that they are in accordance with the approved construction documents. Therefore, the owner should resist the temptation to become needlessly involved in this process and thereby create another step in a critical project control process.

Safety Oversight

In DB projects, safety runs hand-in-hand with constructability and, as a result, the safety control process must begin during design and carry through to the final acceptance of the completed project. Thus, as the constructor provides input to the design, it should ensure that its safety personnel are involved in the continuous constructability review process, because a safe project is easier to build than one whose design is inherently dangerous. Integrating safety planning into the constructability review process creates an environment where all the design-builder's team members must remain aware of the consequences of their decisions on safety. For example, deciding to use precast concrete structural members on a tall building removes the hazard to workers of having to build formwork for cast-in-place concrete members many feet above the ground. It may have other constructability benefits, but the various impacts of each design decision need to be evaluated by the DB team rather than merely by the specific designer who draws up the plans for that feature of work.

Thus, it can be seen from the above discussion that successful DB project administration depends not only on the plans, policies, and processes executed during the project, but, more importantly, is also highly dependent on the actions of the DB team members themselves.

Trust in Design-Build Project Administration

The emergence of DB contracting on the national scene has certainly been controversial. The emotions associated with the paradigm shift required to implement it have run high. When DB contracting emerged in the late 1980s, its detractors consisted primarily of the professional societies associated with the design industry who argued that the use of DB would inevitably degrade the ultimate quality of the constructed product by compromising the integrity of the design process. This fear was expressed in the National Society of Professional Engineers Position Statement No. 1726, which said:

> Design decisions may be determined or inappropriately influenced by team members other than the designer. This is more likely to occur when a non-designer is the lead on the design-build team. The leader may pressure designers to reduce self-imposed quality criteria or design standards to minimum levels in order to maximize profit. (NSPE 1995)

Another DB project administration issue deals with the appropriate distribution of responsibility for quality management in a DB project. To effectively transfer design liability to the design-builder, an owner must also transfer many of the quality assurance responsibilities. This leads to a concern that "The fox may be guarding the hen house," which ties back to the NSPE statement above. A study by Ernzen and Feeney of the Arizona DOT's DB program (appropriately titled "Contractor-Led Quality Control and Quality Assurance Plus Design-Build: Who Is Watching the Quality?") addressed this concern directly by comparing project quality assurance (QA) test data on a DB project where the design-builder had been assigned the responsibility for QA, with data from a similar project delivered by traditional means. It found the following:

> Analysis of the data shows that despite a highly compressed schedule, the quality of the material on the project exceeded the project specifications and was similar to the quality of work completed for the state under traditional contracting methods with an Arizona DOT-operated quality assurance program. (Ernzen and Feeney 2002)

Additionally, in public policy as well as during project administration, perceptions are often as important as facts. Legislative action is heavily influenced by perceptions and, as previously discussed, implementation of DB for public projects has had to overcome the perceptions that DB project delivery would result in an inherently poor quality and possibly unsafe final product because the designer's fiduciary loyalty has been moved to the builder's team. One report on DB implementation classifies perceptions as "barriers to broad acceptance" (Byrd and Grant 1993). An interesting discussion of the issue of perceptions creating a barrier to implementing DB was published by Tom Nicholson in 2005. While it is specifically directed at architectural projects, its content applies directly to all types of projects. The article states, "Architects have groomed a cultural perception that builders can't be trusted" and, as a result, participating in a DB project must be inherently unethical. The author goes on to state, "That perception [that DB is unethical] subsequently contributed to many bidding and contracting laws that made design-build cumbersome or impossible in the U.S." (Nicholson 2005). This perception is contradicted by legislation that specifically authorizes the use of DB on all types of projects across the country. Nevertheless, the perception is stubbornly persistent.

From the above discussion, it becomes evident that a successful DB project administration process will recognize that developing and maintaining a trusting relationship between the owner and the design-builder is paramount. The sophisticated owner will establish its DB design and construction administration processes with this idea in mind and seek opportunities throughout the project's life to foster open communications and continuous improvement. Owners who are trying DB project delivery for the first time must recognize that their personnel and their policies spring from the DBB "catch and punish" approach to project administration. This will be especially true for their quality management systems.

As a result, this book contains two complete chapters that describe the intricacies of design and construction quality management based on the results of research by Gransberg et al. (2007).

The Design-Build Project Team

A focus of our first DB book was on how the owner assembled the proper team to develop the project's scope and then advertised and awarded the DB contract. It also spoke to the design-builder's concurrent actions of developing a design and construction team whose qualifications and past experiences would not only match the owner's requirements but would also win the project. After contract award, the major activity is to weld together the owner's team and the design-builder's team in a manner that facilitates the expeditious completion of the DB project. In an ideal world, the owner's team that developed and awarded the DB project would stay with the project through final acceptance. While this may not be possible, it is extremely important that the institutional knowledge with regard to the project be preserved throughout the project execution period. This is especially critical when it comes to interpreting the design and performance criteria that were used in the RFP during design reviews and construction inspections. Performance criteria are open-ended by definition. Thus, trusting someone who did not develop the criteria to properly interpret them during design and construction is a recipe for dispute.

The same theory applies to the design-builder's team. The group that prepared the proposal would ideally be involved in completing the project. During proposal preparation, both the designer and the builder will generate design assumptions upon which the price proposal will be predicated. It is important that those assumptions be realized, if possible, to ensure that the final as-designed project matches the as-bid project design assumptions and conceptual quantities of work. The design-builder will have also made assumptions about construction means and methods when preparing its proposal, realizing that these are every bit as important as achieving the design assumptions. Often the design-builder has less flexibility about who must work on the project than the owner because of the qualifications evaluation that normally is part of the DB proposal evaluation process.

To establish a climate of trust in the project, it is important that the design-builder ensures that those key personnel whose résumés got the firm selected for the project actually show up on the first day. Once again, the impression of "bait and switch" can poison the relationship between the design-builder and the owner at the outset. Obviously, many things, such as early retirement and job-hopping, can occur between the time a design-builder submits its qualifications and the time it gets the Notice to Proceed with a new DB project that would prevent some of the initial team members from participating in the actual project. If this happens, it is extremely important that the design-builder notify the owner immediately and then make heroic efforts to secure a replacement that the *owner* believes is equal to or better than original person. On the other hand, the owner who delays the award of a DB project should expect the loss of some key personnel because it is not realistic to think that today's design and construction companies can allow their best people to sit idly waiting for a decision by a single client.

The Owner's Design-Build Team

The public owner's team must include all the administrative and technical expertise necessary to allow it to satisfy its responsibilities to the public for delivering a high-quality, safe facility. These same arguments can be made for private owners and developers that will own or sell a facility after its completion. The team typically consists of the following entities:

- Project Management: the personnel or consultants who will control the project on behalf of the owner.
- Procurement
  - Contracting
  - Real estate/right of way
  - Legal counsel
- Design: either internal personnel, consultants, or a combination
  - Engineering
  - Architectural
  - Environmental
  - Special technical consultants
- Construction: either internal personnel, consultants, or a combination
  - Engineering
  - Quality management
  - Safety
- Facility End User
  - Operations
  - Maintenance
  - Tenant organization
- Public Affairs or Marketing: a group responsible for coordinating with both third-party stakeholders.

A private owner will have similar functions on its team, but there will probably tend to be a larger number of consultants rather than internal employees. Additionally, the private owner's project objectives are often financial rather than statutory; the reverse is true for public owners. Thus, the private owner's team may also have members whose expertise is needed to ensure the DB project achieves its pro forma financial requirements, such as a minimum rate of return on the capital investment. Finally, a private owner without any technical design/construction expertise may choose to retain a consultant, such as a construction manager, to manage the DB contract as the owner's agent. In this case, the consultant will need to bring the above list of capabilities to the project.

Owner's Project Management Team

This group is charged with the cradle-to-grave management of the DB project. Ideally, the key members of this group will have been involved in the development of the project scope and its RFQ/RFP, as well as have participated in the DB proposal evaluation process that led to the best-value award decision. This group will lead the project team throughout the project administration process. Its major responsibility is to ensure that, as the details of the project's final form are developed, they conform to the original intent portrayed in the project's scope of work and that all salient objectives for the project's execution are achieved. An example of this type of objective would be to minimize disruption to the traveling public during construction of a DB urban freeway expansion project.

This group often has the authority to authorize progress payments to the design-builder for design and construction. This entity will also have the authority to make changes to the contract scope of work if and when they become necessary. Ultimately, this group is directly responsible to the owner's senior management for the success of the project. This group will normally be made up of senior technical and administrative personnel, and its charter will be to keep the global perspective of the project in focus as the details are created.

Owner's Procurement Team

Depending on the project and the owner's internal organizational structure, the procurement team may have either a very small or a fairly large part to play during project administration. The U.S. federal government, for instance, has evolved a very complex structure around a position called the "contracting officer." This position is not normally filled with a technical person. Rather, it is a purely administrative position that receives advice and guidance from the technical experts on the owner's design and construction teams. No change of awarded DB contract scope can be made without the contracting officer's approval (FAR 1997). Thus, a federal DB project will see the active, continuous involvement of the federal owner's procurement team during project administration. Private owners will usually assign this type of authority to its project management team and, as a result, the procurement team will disappear upon contract award. Public and private owners will also rely on legal staff to draft the procurement documents. In the final analysis, this team's major responsibility is to ensure that the project is delivered in a manner that conforms to all applicable legal and regulatory standards, as well as to deal with any disputes that may arise.

Owner's Design Team

The owner's design team must have the requisite technical expertise to be able to adequately review the design submittals produced by the design-builder as well as to knowledgeably interpret the performance criteria contained in the DB scope of work. Ideally, the members of this team will be the same people who developed the project's performance criteria and thus will intimately understand their technical intent. This group will form the nucleus of the owner's design submittal review team and will normally be the people who implement the design quality management tasks for which public owners are statutorily responsible. This will ensure that the conditions established by project permits as well as design criteria and building codes are satisfied by the design-builder. Often, the owner will supplement this group with special technical consultants who have unique expertise in various features of the project's scope. Many state DOTs retain a general engineering consultant to augment their internal design staff on fast-moving DB projects (Gransberg et al. 2007). This makes sense in that the design review activities are always on the project's critical path, and it behooves the owner to ensure that all design reviews are completed as expeditiously as possible. It also reinforces the partnering agreement with the design-builder by tangibly demonstrating the owner's commitment to not only maintaining a cordial working relationship but also assisting the design-builder in hitting its most critical project milestones. It is important that the owner not reassign this team after the final design is released for construction because they will be needed to assist the owner's construction team with field changes as well as final acceptance activities. Additionally, experienced owners have found that it is important to appoint one individual on this team as the single point of contact for all design issues, thereby ensuring that all design decisions flow through a single point and are thus properly coordinated with other design disciplines on the team.

Owner's Construction Team

This group will represent the owner's interests during construction. Once again, this team should be composed of individuals with the requisite technical and experiential background to be able to authoritatively interface with the design-builder's field staff, as well as able to handle design issues that may appear during construction. Because the owner is no longer the DOR's direct client, and the designer and builder are on the same side of the contract, the chain of paperwork that is associated with making design changes during construction is abbreviated. The design-builder can make design decisions verbally in the field as long as they are properly documented in the as-built drawings and specifications. Thus, the owner's construction staff must often be augmented with design personnel who have the necessary expertise to evaluate the impact of design changes made in the field for the convenience of the constructor. The team may need special expertise to oversee the various construction engineering and quality management tasks that will occur during the course of construction.

Facility End User

The role of this entity on the owner's DB team cannot be overemphasized. This is the organization that must ultimately take over the DB project after it is constructed and put it to its intended use. This group should have made extensive input to the project's scope of work during RFP development. After project award, this group needs to remain involved, expressing preferences and furnishing input regarding operability and maintainability issues during design and construction. In the final analysis, if this entity is not satisfied with the final constructed product, the DB project is a failure.

Owner's Public Affairs/Marketing Team

This group is normally responsible for coordinating with entities outside the normal project team. This includes third-party stakeholders such as utility companies and railroads, the general public, and the news media. It is not uncommon to find this group assigned to the project management team because of the importance of their efforts. Their main objective is to take care of the external issues for the project team, allowing the other members to focus on getting the project designed and built without having to be distracted by short-term, low-impact crises originating outside the project itself. In private projects, the owner often employs a marketing team to help advertise the project and liaise with groups directly affected by construction.

The Design-Builder's Team

The design-builder's team includes all the people and organizations that were promised in the project's winning proposal, plus those who were not specifically identified but from whom necessary service is required to complete the project in accordance with the DB contract. It typically consists of the following entities:

- Project Management: the personnel or consultants who will control the project on behalf of the design-builder
- Design:
  - — Design administrator
  - — Designer-of-Record
  - — Design subconsultants
  - — Special technical consultants
- Construction:
  - — Construction manager
  - — Trade contractors
  - — Material/equipment suppliers
  - — Cost engineering/project controls
  - — Quality management
  - — Construction engineering
  - — Safety
- Support Staff:
  - — Legal counsel
  - — Logistics
  - — Real estate/right of way
  - — Public affairs.

Design-Builder's Project Management Team

The design-builder's project management team contains the senior leadership necessary to execute the project in accordance with the contract. The team should be the same people who prepared the winning proposal and who have an intimate knowledge of the assumptions, allowances, and contingencies that were used to arrive at the proposed price. Additionally, the team understands the design-builder's business objectives as well as the basis for the target profit margin used in the proposal. In the case where the design-builder is actually a consortium of several companies that was formed to compete for the given project, the project management team will contain members from each equity partner to the consortium. It will have the authority to obligate the design-builder in contracts and to authorize the payment of subcontractors and material suppliers. It will also have the authority to make changes in the agreements that were consummated in conjunction with the project. It is important that this group have a clear leader to whom the final decision in all business matters relating to the DB project can be referred. This individual will serve as the design-builder's point person in all dealings with the owner. While other members of the team will be able to deal with members of the owner's team in routine project management matters, it is key to the project's success that the designated leader be kept informed of these dealings, and that when conflicts arise the leader become involved in their resolution.

Design-Builder's Design Team

The design-builder's design team is led by the design administrator. This is an individual with strong design credentials whose primary duty is to manage the design process and to coordinate it with the construction process. This person is different from the DOR in that he or she has no design production responsibilities. This position acts as a single point of contact for the owner on all design issues. This individual is personally responsible for seeing that these issues are resolved in a manner that facilitates achievement of the project schedule and that the final design conforms to the budget portrayed in the price proposal.

The DOR leads the design production team itself. This individual is the single point of responsibility for all design decisions and design products for the design-builder's design team, and may supply the required professional liability insurance. The DOR reviews, coordinates, deconflicts, and approves for construction all design and extensions of design produced by all members of the design-builder's design team. The DOR may also lead the design quality control effort as appropriate.

The design team is rounded out with all the necessary design subconsultants and technical specialists necessary to complete the project's final design. The team will also remain intact throughout construction, performing construction quality assurance tasks as required by the contract. In some fast-moving DB projects, the owner has required that the design team literally co-locate with the construction team and conduct the DB design effort at or near the project site.

Design-Builder's Construction Team

The construction team looks much like any contractor's construction team and is made up of those forces necessary to complete the work using the means and methods assumed in the proposal. However, in DB, this team must be prepared to work with less than a complete design for most features of work. Additionally, the trade subcontractors and materials suppliers must be prepared to develop and furnish typical construction submittals at a much earlier stage, usually during the design phase, to allow the design-builder to leverage the DB process by minimizing the internal construction administration effort. Another difference is the continuous involvement of the construction cost engineering group in the design effort, making sure that the final design does not exceed the cost parameters contained in the price proposal. This must be done hand-in-hand with the DOR through the offices of the design administrator so that lost design effort by the design team is also minimized. The construction quality management team will take its form from the contractual allocation of quality control and quality assurance responsibilities articulated by the owner in the DB contract.

Design-Builder's Support Staff

This group furnishes the backbone that gets the thorny details of the project's administration and logistics accomplished. It will always coordinate the procurement of materials and the efforts of the trade subcontractors. If the project includes the procurement of real estate or right-of-ways, this group will take the lead at an early stage in the project to find the land upon which the project will ultimately be built. Many large DB projects in the transportation and public works sectors also require the design-builder to retain a public relations firm to coordinate community outreach, public information, and news media coverage of project progress.

Project Team Summary

DB project delivery requires both the owner and the design-builder to assemble teams that match the DB project itself. It is also helpful to have clear understanding and lines of authority between the owner and the DB teams. Additionally, because DB requires the fixing of project cost before design completion and because compressing the project's schedule is normally one of the project objectives, it is important that the team members are not only competent but also flexible in their ability to react to the challenges associated with this project delivery method. When assigning resources to the DB project team, both the owner and the design-builder need to remember that the scope is defined by performance criteria whose proper interpretation requires the mature professional judgment that springs from relevant past experience. When putting together a DB project team, the owner must be brutally honest regarding the abilities of its organization to adapt to this project delivery method. If it has no previous DB experience, it would be wise to retain a DB consultant to assist the team in making the transition from DBB to

DB. Every DB project has one specific project goal that it must achieve, and the owner had that goal in mind when it made the decision to select DB project delivery. To achieve that preeminent goal, both the owner and the design-builder must have the right teams.

Identifying the Preeminent Factor for Using Design-Build

Perhaps the greatest secret to DB project administration success lies in the owner and the design-builder always keeping in mind the owner's major reason for selecting the DB project delivery method and then using that factor as the yardstick against which to assess project decisions. At project inception, the owner will have identified the various aspects of the project that must be achieved to meet its requirements. Generally, these will fall into the categories of cost, schedule, and quality as defined by the technical design. Of these three factors, the project will normally have one that is the most important for this project's ultimate success. This will be called the "preeminent factor," and it is the one for which the owner is prepared to sacrifice pieces of the other two so that this factor will be achieved.

An excellent example of this comes from the Utah DOT's I-15 project in Salt Lake City that was completed in 2002 (Postma et al. 2002). While UDOT obviously had a fixed budget and certain standards to maintain with regard to quality, schedule was the preeminent factor because the project had to be finished prior to the start of the 2002 Winter Olympics. That factor was the reason why UDOT selected DB project delivery. In this case, the owner could not complete the necessary work using the traditional process in time to meet the deadline. Thus, a project whose preeminent factor was schedule had to be designed and built seeking every opportunity to compress the project delivery period. If there were two possible technical design solutions to a given feature work, the design-builder with the owner's support would select the option that could be designed and built in the shortest amount of time. As a result, it was vital to project success that every member of every team on both sides of the DB contract was aware of the project's preeminent factor so that they could perform their duties in a manner that supported its achievement.

Examples of other preeminent factors can be found in published project goals in RFPs. Published project goals relating to highway construction are provided in Table 1-2. The project goals in Table 1-2 vary in style and emphasis due to the unique project needs, but they all help define the agencies' requirements in terms of schedule, cost, quality, aesthetics, and end user requirements. The application of these goals throughout the project administration process will help owners and design-builders successfully administer their project and avoid potential disputes.

It is possible to award a DB contract where there is no price competition. This is called "Fixed Price-Best Proposal" (Gransberg and Molenaar 2004). In this particular case, the owner has a fixed budget and is advertising to see how much "bang" it can get for its "buck." When a DB project is awarded using this method, the preeminent factor is cost. Therefore, both the owner and the design-builder

Table 1-2 Examples of Design-Build Project Goals

Agency	Project	Project Goals (in descending order of importance)
Colorado DOT	Colorado Springs Metro Interstate Expansion Project	• Maximize capacity and mobility improvements in the corridor within the program budget of approximately $150 million • Minimize inconvenience to the public during construction • Provide a quality project • Complete by the end of calendar year 2008 • Provide a visually pleasing final product
	Southeast Corridor Multi-Modal Project (TREX)	• Minimize inconvenience to the community, motorists, and the public • Meet or beat the total program budget • Provide for a quality project • Meet or beat the schedule of June 30, 2008
New Mexico DOT	US-70 Hondo Valley	• Cost not to exceed budget • High quality, safe, aesthetic, environmentally responsible, durable and maintainable project • Contract awarded and signed by June 2002 • Project completion not later than September 25, 2004 • Valid basis for evaluation of DB delivery system
South Dakota DOT	Interstate 229	• Timely completion • Quality design and construction • Reasonable cost
Washington DOT	Everett HOV Design-Build Project	• Deliver the project within budget • Achieve substantial completion by October 1, 2007 or sooner • Achieve quality of design and construction equal or better than traditional DBB • Provide a safe construction site for workers and the traveling public • Meet or exceed environmental requirements and expectations with no permit violations • Foster confidence with the environmental permitting community in the DB process • Manage traffic to minimize disruption and inconvenience to the public during construction • Maintain community support during design and construction
	I-405 Kirkland Stage I	• Quality of design and construction (on time within budget) • Environmental compliance and innovation • Maintenance of traffic • Public information and community involvement

must approach such a project with the idea that the *scope* is now variable, rather than the *price*. Thus, if an unexpected condition arises where the design-builder will be due extra compensation that would exceed the owner's budget, it is incumbent upon the design-builder to seek options that not only mitigate the cost of the unforeseen condition but also reduce costs elsewhere in the project to allow the owner to achieve its budget. In the private sector, it is not uncommon to associate a guaranteed maximum price (GMP) with these types of projects.

Finally, there are those projects where quality, as expressed by the technical design and/or the service it provides, is the preeminent factor. This would be the case in a building whose purpose is to serve as a piece of monumental architecture. This is also the case in an industrial process facility whose purpose is to produce an industrial product, and the DB project is merely the initial capital investment in the process. In this case, an unforeseen spike in the price of construction materials will need to be absorbed, or financing to reimburse the design-builder for the cost of an unforeseen site condition will need to be found to preserve the technical quality of the DB project.

Clearly identifying, communicating, and discussing the ramifications of the DB project's preeminent factor among both the owner's and the design-builder's project teams will pay dividends during project administration. It furnishes a basis for decision making. It also provides a reason to generate and evaluate alternatives for design and construction. Finally, it furnishes a common ground where all the players can plan and execute their individual responsibilities. In fact, it can become the theme around which the project's partnering and team-building efforts can be centered.

Partnering the Design-Build Contract after Award

An important factor in achieving a high-quality DB project is free and open communication between all parties during the design phase (Beard et al. 2001). DBIA's *Manual of Policy Statements* states: "DBIA advocates both formal and informal project partnering and considers the partnering philosophy to be at the foundation of design-build delivery" (DBIA 1998). Partnering is a concept that seeks to bring together the various parties to a contract in a manner that creates an environment of open communication and trust. It has been used successfully in DB projects since the late 1980s.

The term "partnering" evokes different meanings in different sectors of the engineering and construction industry. Among the designers and builders of privately financed projects, partnering is a strategic relationship that is developed for relatively long periods and for multiple projects. These strategic partnerships garner many advantages for their members. The main one is the development of a thorough understanding of the partners' motivations, trustworthiness, and means of communication. This understanding allows one partner to gauge the other partner's potential reactions to an impending crisis. It also encourages the honest sharing of bad news in a timely manner, which permits joint action to avert or minimize the impact of the crisis on the successful completion of the project in question.

Partnering in the Public versus Private Context

Private strategic partnerships have an advantage over their counterparts in the public sector in that private entities are relatively free from regulation of the form and substance of their internal operational activities and contractual relationships. Public agencies must answer to lawmakers, regulators, and the public alike. Thus, the freedom for public agencies to develop long-standing, strategic partnerships with private organizations is greatly diminished if not eliminated altogether. As a result, public agencies have traditionally focused their partnering activities on single-project team-building seminars. However, it should be noted that DB project delivery is used in public–private partnership (PPP) projects, and recently enacted public laws such as the FHWA Special Experimental Program 15 (FHWA 2007) now permit many public agencies to engage in the long-term strategic partnerships that formerly were confined to the private sector. The remaining discussion of partnering in the DB context will be confined to the single-project version described below.

The literature shows that the growth of project partnering is directly related to the growth in claims and litigation regarding construction contracts throughout the United States (Kubal 1994). In the late 1980s, the USACE led the way for public agencies to begin using this new business practice as a means to avoid disputes and consequently reduce the ultimate cost of delivering public facilities. To verify the success of partnering, projects must be measured. One pitfall in past efforts to measure partnering performance involved the collection and interpretation of statistics regarding partnering. In public agencies, there was a tendency to credit partnering for project successes even when there was no tangible evidence of any improvement over the status quo. This was caused by the intense personal investment public project managers and contractors made during partnering sessions. There is no doubt that enhanced communication greatly improves a project's management/dispute resolution environment. However, most early studies of the process failed to identify significant benefits that can be directly attributed to partnering programs. The first study of the Texas DOT's (TxDOT) partnering program (Grajek 1995) found that partnered projects finished an average of 13.73% ahead of schedule as compared to non-partnered projects (9.68%). While this appears to show some partnering impact, the fact that most projects finish ahead of the contract completion date indicates that TxDOT is generally conservative in establishing contract duration. While there is nothing fundamentally wrong with this policy, it makes interpretation of actual performance data difficult with regard to schedule. A study of the Ohio DOT's partnering program (Chapin 1994) addressed only cost growth and found that 20 partnered projects had a 1.00% cost growth as compared to 123 non-partnered projects which averaged a 4.03% cost growth. Although the partnered sample size was small, it does seem to suggest potential benefits for partnering on DBB projects.

USACE found that partnering is most valuable on projects with tight schedules where techniques such as issue escalation and open communication tend to enhance the efficiency of critical decision making. The constructor is allowed the maximum amount of time to react to scope changes and retain satisfactory progress. Additionally, change order time extensions are much more important to a constructor on a project with a tight schedule than on one that has greater schedule flexibility (Kubal 1994). Thus, the constructor will be more liable to formalize a dispute over a time extension on a project that has a tight schedule rather than when there is greater schedule flexibility (Kane 1992). This fact further blurs the validity of the apparent schedule improvement on partnered TxDOT projects.

The above discussion is not meant to cast doubts on the validity of the partnering process but, rather, to indicate the importance of understanding the dynamics of the process that produces the contract performance data. Studies done on USACE and Naval Facilities Command (NAVFAC) projects were confined to competitively bid, firm-fixed-price projects (Pina 1993; Schmader and von Rosenvinge 1994; Weston and Gibson 1993). Since the time frame of those studies, "best value selection" has been implemented on a broad scale by USACE and NAVFAC. Best value selection removes the requirement to award to the low bidder and has changed the dynamic under which partnering was developed in the federal government (Ellicott and Gransberg 1997). This approach shows much promise in that strategic relationships that produce positive outcomes for both parties are able to reap benefits from the synergy of repetition.

A later study on a much larger sample of TxDOT's early DBB partnering program found that partnered projects outperformed non-partnered projects in every category in which they were compared and provided an effective means to control both cost and time growth (Gransberg et al. 1999). Additionally, the field survey showed that 60% of TxDOT personnel and 85% of constructor personnel felt that implementing partnering improved the quality of the project as well (Gransberg et al. 1998). A study of NAVFAC's DB partnering process produced similar results and dramatically demonstrated the benefits of applying partnering principles to DB projects (Allen et al. 2002). Finally, a study of the Arizona DOT's partnering efforts on a major DB project in Phoenix found that partnering DB transportation projects was a natural fit, stating:

> Design-build by its nature lends itself to the partnering concept. The partnering concept ideas of increased communication, alignment of goals, and development of a dispute resolution system fit perfectly with design-build's overarching theme of single-point responsibility for the owner. Increased pressure because of schedule compression typical of most design-build projects makes partnering a vital necessity. (Ernzen et al. 2000)

Because most formal partnering programs begin shortly after contract award, it is logical to include this discussion of the benefits of partnering in the first chapter of this book. It should be noted that in many cases, actual partnering begins during the design-builder's team-building period that accompanies proposal preparation during the procurement phase for the designers and constructors on the design-builder's team. Additionally, many owners apply partnering principles to developing strong working relationships with external stakeholders, such as public environmental agencies, political entities, major property owners that will be affected during the project, special consultants, and so forth. These procurement-phase partnering efforts often involve internal lawyers, engineering discipline areas, and procurement personnel, and can have the same impact on the project as the DB team partnering prior to contract award.

Notwithstanding the pre-award internal partnering opportunities for the owner's and the design-builder's teams, the first opportunity to extend that partnering effort to the owner's and design-builder's project personnel occurs upon contract award. One goal of the initial partnering meeting should be to create the conduits of communication that are necessary to transmit critical design information and owner preferences to the design-builder's design team. Another goal is to develop the structure for ensuring that not only the designers on the owner's and designer's team are aware of critical information, but also that the constructors are included in the information loop to ensure that the level of constructability upon which the price proposal was established is not compromised during the design phase.

Design-Build Partnering Communications Protocol

"Open communications is the key to any partnering process" (Ernzen et al. 2000). The DB team for the Arizona DOT's DB project described earlier developed a protocol based on a series of weekly joint meetings after the initial partnering workshop to foster an environment of free-flowing communications. Figure 1-5 displays the content and outcomes of those routine meetings, as well as the attendees. Of particular importance for this synthesis is the fact that the design-builder's primary design and construction quality managers were prominently involved in each of the sessions.

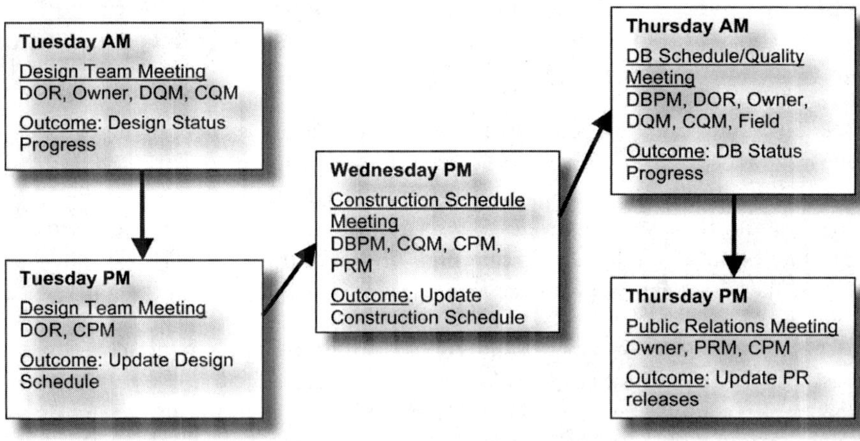

Figure 1-5 Arizona DOT I-17 design-build project partnering communications protocol. DOR, designer-of-record; DQM, design quality manager; CQM, construction quality manager; DBPM, design-build project manager; CPM, construction project manager; PRM, public relations manager.
Source: Adapted from Ernzen et al. (2000).

One can see in Fig. 1-5 that the outcomes of one meeting drive the subject of the next meeting, starting with design status and eventually ending up with a public relations release that updates the traveling public and other third-party stakeholders on developments of interest in the project. This commitment to partnering in DB project execution creates a mechanism to not only ensure that design quality issues are addressed as they are encountered, but also to extend the design decisions into the construction phase and evaluate their impact on constructability as well as resultant construction quality. The regular and routine involvement of the owner's project team members provides a point at which owner quality management activities, such as over-the-shoulder design reviews and other quality management tasks, can be undertaken if the owner has decided to assign design and construction QA and QC responsibilities to the design-builder. Some owners enhance this effort by requiring that the design-builder's design team be co-located with the construction team and, at times, the owner's DB project personnel.

Design-Build Success Factors

Some common themes for successful design-build projects are presented throughout this book. The reader is encouraged to keep the following items in mind as they read the text:

- DB requires a higher level of trust and partnering.
- DB requires the owner to develop definitive, functionally driven performance criteria.
- DB requires a cultural shift away from the DBB mentality.
- Remember who owns the details of the design.
- Get the team together early and keep it together.
- The DB contract is a "construction" contract that also covers design.
- Constructors have been doing "design" all along in the form of shop drawings and submittals.
- DB is a scope-driven endeavor.

References

Allen, L. N., Gransberg, D. D., and Molenaar, K. R. (2002). "Partnering successful design-build contracts." *Military Engineer,* 93(609), 23–24.

Beard, J. L., Loulakis, M. C., and Wundram, E. C. (2001). *Design build planning through development,* McGraw-Hill, New York.

Byrd, L. G., and Grant, A. A. (1993). "Prerequisites for a successful design/build/warranty highway construction contract." *USDOT Federal Highway Administration,* <http://www.fhwa.dot.gov/programadmin/contracts/byrd.cfm> [Feb. 4, 2007].

Chapin, L. T. (1994). "Evaluation of partnering on Ohio Department of Transportation projects." *Ohio Department of Transportation Report FHWA/OH-94/022,* Ohio Department of Transportation, Columbus, Ohio.

Design-Build Institute of America (DBIA). (2005). "Non-residential construction in the United States." *Design-Build Institute of America Dateline,* May, 7–8.

DBIA. (2000a). "Market trends, water/wastewater outsourcing revenues jump in 1999." *Design-Build Institute of America Dateline,* May, 1-2.

DBIA. (2000b). *Guide to the federal marketplace.* Design-Build Institute of America, Washington, D.C.

DBIA. (1998) *DBIA manual of policy statements,* Design-Build Institute of America, Washington, D.C.

Eby, M. (2005). "The inevitability of design/build." *EC&M,* <http://ecmweb.com/ar/viewpoint_0605/> [Feb. 2, 2007].

Ellicott, M. A., and Gransberg, D. D. (1997). "Best value contracting criteria." *AACEI Cost Eng.*, (39), 31-34.

Ellis, R. D., Herbsman, Z., and Kumar, A. (1991). "Evaluation of the FDOT design/build program." *Final Report, Submitted to Florida Department of Transportation, State Project No. 99700-7543-010,* Department of Civil Engineering, University of Florida, Gainesville, Fla.

Ernzen, J., and Feeney, T. (2002). "Contractor-led quality control and quality assurance plus design-build: Who is watching the quality?" *Transp. Res. Re.: J. Transp. Res. Board,* 1813, 253–259.

Ernzen, J., Murdough, G., and Drecksel, D. (2000). "Partnering on a design-build project: Making the three-way love affair work." *Transp. Res. Rec.: J. Transp. Res. Board,* 1712, 202-212.

Federal Acquisition Regulation (FAR). (1997). "Two-phase design build selection." *Federal Register,* Vol. 62, No. 1, 62 FR 271.

Federal Highway Administration (FHWA). (2007). "SEP-15 Program—Public private partnerships." *USDOT Federal Highway Administration,* <http://www.fhwa.dot.gov/ipd/p3/tools_programs/sep15.htm> [March 10, 2010].

FHWA. (2006). "Study to congress on the effectiveness of design-build project delivery relating to the Federal-Aid Highway Program—Final report." *USDOT Federal Highway Administration,* <http://www.fhwa.dot.gov/reports/designbuild/designbuild0.htm> [Aug. 30, 2006].

FHWA. (2002). "FHWA design-build rule." Published in the Dec. 10, 2002 *Federal Register,* <http://www.fhwa.dot.gov/programadmin/contracts/dbfinal.cfm> [Dec. 1, 2006].

Grajek, K. M. (1995). "Partnered project performance in the Texas Department of Transportation." Master's thesis, University of Texas, Austin, Tex.

Gransberg, D. D., and Molenaar, K. R. (2007). *The impact of design-build on the public workforce.* Keston Institute for Public Finance and Infrastructure Policy, University of Southern California, Los Angeles, Calif.

Gransberg, D. D., Molenaar, K. R., and Datin, J. N. (2007). "Quality assurance in design-build projects." *NCHRP Synthesis 38-01,* Transportation Research Board of the National Academies, Washington, D.C.

Gransberg, D. D., Koch, J. E., and Molenaar, K. R. (2006). *Preparing for design-build projects: A primer for owners, engineers, and contractors,* ASCE Press, Reston, Va.

Gransberg, D. D., and Molenaar, K. R. (2004). "Analysis of owner's design and construction quality management approaches in design-build projects." *ASCE J. Mgmt Eng.*, 20, 162.

Gransberg, D. D., and Villarreal, M. E. (2002). "Construction project performance metrics." *2002 AACE Trans.*, AACE International, Morgantown, W.V.

Gransberg, D. D., Dillon, W. D., Reynolds, H. L., and Boyd, J. (1999). "Quantitative analysis of partnered project performance." *ASCE J. Const. Eng. Mgmt.*, 125(3), 161–166.

Gransberg, D. D., Reynolds, H. D., Gilman, A., and Gokdogan, G. (1998). "Evaluation of the Texas Department of Transportation Partnering Plus Program." *Technical Report No. TX98/0-1729-S*, Texas Department of Transportation, Austin, Tex.

Hancher, D., and Werkmeister, R. (2001). "Managing change in state departments of transportation: Innovations in private involvement in project delivery." *Transportation Research Board of the National Academies*, <http://onlinepubs.trb.org/onlinepubs/nchrp/nchrp_w39-1.pdf> [March 10, 2010].

Hicks, R. G., and Epps, J. A. (2000). "Life cycle cost analysis of asphalt-rubber paving materials." *Proc., Asphalt Rubber–2000*, Rubber Pavements Association, Tempe, Ariz., 569–590.

Kane, C. (1992). "Mitigating construction contract disputes." *Public Utilities Fortnightly*, 130(1), 11–12.

Kubal, M. T. (1994). *Engineered quality in construction: Partnering and TQM*. McGraw-Hill, New York.

Loulakis, M., ed. (2003). *Design-build for the public sector*, Wolters Kluwer Law & Business/Aspen Publishers, Inc., New York.

Molenaar, K. R., and Scott, S. (2003). "Examining the performance of design-build in the public sector." *Design-build for the public sector*, M. Loulakis, ed., Wolters Kluwer Law & Business/Aspen Publishers, Inc., New York, 71–112.

Molenaar, K. R., Songer, A. D., and Barash, M. (1999). "Public sector design-build evolution and performance." *ASCE J. Mgmt. Eng.*, 9(2), 54-62.

National Society of Professional Engineers (NSPE). (1995). "Design/build in the public sector." *NSPE Board of Directors, Position Statement No. 1726*, <http://www.nspe.org/GovernmentRelations/TakeAction/PositionStatements/ps_des_bld_pub_sec.html> [March 10, 2010].

Nicholson, T. (2005). "AIA teaching architects to lead design-build teams." *Design-Build Magazine*, <http://designbuild.construction.com/features/archive/2005/0506_feature2.asp> [Feb. 4, 2007].

Pantazides, L., (2005). "Managing quality on transportation mega projects." *Proc., ASQ World Conference on Quality and Improvement*, Vol. 59, American Society for Quality (ASQ), Milwaukee, Wisc., pp. 289–297.

Peters, M. (2003). "An important project." *USDOT Federal Highway Administration*, <http://www.fhwa.dot.gov/pressroom/re031021.htm> [Sept. 16, 2006].

Pina, B. D. (1993). "Partnering at the Naval Facilities Engineering Command." Master's thesis, Purdue University, West Lafayette, Ind.

Postma, S. E., Cisneros, R., Roberts, J., Wilkinson, R., Clevenger, J., and Eastwood, A. (2002). "I-15 corridor reconstruction project design/build evaluation 2001 annual report." *Report No. UT-02.11*, Utah Department of Transportation Research Division, Salt Lake City, Utah.

Science Applications International Corporation (SAIC). (2003). "Outsourcing of state DOT capital program delivery functions." *Transportation Research Board of the National Academies*, <http://trb.org/publications/nchrp/nchrp_w59.pdf> [March 20, 2010].

Schmader, W. J., and von Rosenvinge, T., IV. (1994). "Partnered project performance in the Naval Facilities Engineering Command." *Proj. Mgmt. J.*, 26(3), 39–48.

Songer, A. D., and Molenaar, K. R. (1996). "Selecting design-build: Private and public sector owner attitudes." *ASCE J. Eng. Mgmt.*, 12(6), 47–53.

U.S. Army Corps of Engineers (USACE). (2006). "MILCON transformation model RFP." *U.S. Army Corps of Engineers*, <http://www.wbdg.org/ccb/ARMYCOE/COEECB/ecb_2006_13.pdf> [March 10, 2010].

Washington State Department of Transportation (WSDOT). (2004). *Guidebook for design-build highway project development*, Washington State Department of Transportation, Olympia, Wash.

Weston, D., and Gibson, G. E. (1993). "Partnering: Project performance in the U.S. Army Corps of Engineers." *ASCE J. Mgmt. Engrg.*, 9(4), 410–425.

Zweig-White Research. (2004). "Is design-build due for an increase?" *The Zweig Letter, Issue 551*, <http://www.zweigwhite.com/trends/thezweigletter/index.asp> [Feb. 18, 2007].

TWO

Scheduling Design-Build Projects

In 1996 the University of Colorado conducted an owner survey to identify the factors that are used by both public and private owners in selecting design-build (DB) project delivery (Songer and Molenaar 1996). The essence of this research was to quantify owners' opinions as to the perceived benefits of DB. That study found that there is one primary reason why owners select DB: *to shorten project delivery period duration*. A 2003 study for the National Cooperative Highway Research Program (NCHRP) of the National Academies' Transportation Research Board recorded the perceptions of state department of transportation (DOT) upper managers regarding their perceived benefits of implementing DB. It found those managers to believe that "... design-build projects opened to the public much sooner ... [and] there is a considerable saving to the public in the form of user cost benefits due to significant schedule accelerations" (SAIC 2003, p. v). Finally, in 2004 the Federal Highway Administration (FHWA) published its study on the effectiveness of DB in transportation projects. The study addressed project performance in the areas of cost, schedule, and quality, and it found that "[D]esign-build project delivery reduced the overall duration of their projects by 14 percent ... as compared to design-bid-build project delivery" (FHWA 2004). These two recent studies confirmed the Songer and Molenaar (1996) research conducted more than a decade ago, which found that the primary reason for public owners selecting DB project delivery was to compress the project schedule. This leads to the conclusion that scheduling is a critical element of success in DB project delivery and therefore must be given prime attention for no other reason than to meet the owners' expectations.

There are three situations that drive many owners to select DB to compress the delivery schedule. In the first case, a project that has a postconstruction revenue stream associated with it, such as a toll bridge, a hotel, or a water treatment plant, will increase its profitability and hence its economic viability for every additional day of revenue that can be derived from the early opening of the project. From this perspective, it is easy to understand why gaming companies in Las Vegas have selected DB as the method for building their new casinos. Even less glamorous projects like commercial parking garages or strip malls will incrementally enhance their long-term profitability if their owners are able to open their new facilities days or weeks earlier than would have happened using traditional project delivery. The concept that allows this to happen is permitting construction to begin before design is complete. In essence, the owner is consciously deciding to not wait until the color of the interior paint has been determined before starting to dig the foundations.

The second situation most often occurs in public sector projects where the expiration of funding authority, perhaps at the end of a fiscal year, creates a need to "obligate" the designated funding through contract award before the drop-dead date. In this case, owners select DB when they realize that there is not enough time to award a design contract, complete the design, and then advertise, bid, and award the construction contract. This can also happen in the private sector when the same set of circumstances places the construction finish date beyond the project's required delivery date. Thus, seeking to reduce the schedule by allowing concurrent design and construction activities allows the owner to complete its project in the time available. The I-15 project in Utah is a recent example of this. The Utah DOT had to complete the billion-dollar project before the start of the 2000 Winter Olympics (Postma et al. 2002) and used DB as a way to successfully accelerate both design and construction. In another example, a commercial building project in Oregon used DB where the project's financing depended on awarding the construction contract by a particular date to lock in the interest rate. DB was the only way to award the construction contract on a lump-sum basis before the design was complete and as a result was selected.

The third reason is most keenly felt in the public transportation sector, where the need to increase capacity rapidly while minimizing congestion during construction creates another objective. Timeliness is the paramount factor in transportation infrastructure projects, and nowhere is it more critical than on congested urban freeway projects. User costs of congestion due to delays, increased toxic emissions, and other factors on urban freeways can exceed $10,000 per lane-mile/day (Hicks and Epps 2000). Thus, a typical eight-lane urban freeway would incur a user cost of $80,000 per mile per day, and a hypothetical five-mile urban freeway upgrade project would generate user costs on the order of $400,000/day. Taking this analogy one step further would lead one to determine that the 30-day period normally used to advertise and award a traditional DBB project would cost the commuters $12 million on that five-mile stretch of freeway; thus, DB would generate a significant benefit by delivering a needed upgrade just one month earlier than it would have been delivered using traditional means. While the actual value of user costs is a subject of much debate, there is no argument about the fact that they indeed are incurred, and the 14% average reduction in project duration cited in the FHWA study (2004) would translate to a commensurate 14% reduction in user costs of congestion both during construction and after the improvement is made.

In all of the above situations, the owner is "getting" an early delivery in exchange for "giving up" direct control of the details of design to the design-builder. Therefore, by moving control of both design and construction to a single entity, that entity is able to start construction at the earliest possible point. This does not necessarily imply a "fast-track" project as defined in Chapter 1. In the commercial building example described above, the reason for selecting DB was related to a financial deadline rather than the desire to open the facility as soon as possible. Therefore, the owner ultimately required that the design-builder complete the construction documents and pull all the permits before authorizing construction to begin. Thus, selecting DB does not force the owner to allow the design-builder to proceed as rapidly as possible. A DB contract structured to include both design and construction Notices to Proceed allows the owner to complete the project in much the same manner as a traditional one, while quantifying the construction costs at an early point in the design process and saving the time it takes to advertise and award a construction contract after design is complete.

All of the above discussion is to lead the reader to the understanding that scheduling is particularly critical to DB project success. It also shows the importance of coordinating the scheduling of design products with their corresponding construction features of work. To put it simply, to build a schedule for the design effort that does not inextricably interconnect with the construction is to fail to frame the problem in its appropriate context. Gauer (2006) puts it like this: ". . . on a Design Build project, the design activity is *always* on the critical path. The key challenge for the engineering team is for the design to get ahead of construction at the soonest possible date . . . there is a simple need to design first what the contractor will build first." Hence, there is a clear need to ensure the integration of the DB project's schedule to plan and control the integrated delivery of the project itself.

Analyzing Contract Schedule Requirements

When scheduling a DB project, the first item that must be checked is the contract itself, which will undoubtedly contain contractual schedule factors that originated in the project's Request for Proposal (RFP) and/or the winning proposal. These schedule requirements will embody many of the possible solutions to the owner's project delivery timeline problem. The first type are those that actually define the milestones on the project schedule itself. These are explicit constraints on the time function associated with this project. They are:

- Required delivery date
- Intermediate completion dates
- Construction phasing requirements
- Site availability date

The next type are those that constrain the pace of the project itself. These are more subtle in nature and operation, and the DB team must seek to clearly understand their impact on potential scheduling solutions by the design-builder. Examples of these are:

- Permitting and external design/project review requirements
- Cash flow considerations
- Fast tracking

At the earliest possible moment, the DB scheduling team should ensure that the pace constraints are not in conflict with the contractual schedule requirements. For example, a public owner using fiscal year constrained funding for a multi-year project may have unintentionally created a requirement to phase construction in a manner that involves constructing more product than the owner has the funding authority to reimburse. Also, a project that requires a significant environmental permitting process by an external agency with no incentive to facilitate progress may not be able to realistically achieve specific intermediate milestones through no fault of the project's owner or the design-builder.

The final thing that the scheduling team should do is to make sure that the schedule factors are in harmony with the technical requirements and that no unintentional conflict has been created that would jeopardize timely project completion. An example of this is a project that includes an item of equipment with a long lead time between order and shipping.

Design-Build Contract Scheduling Clauses

As with all other aspects of construction project delivery, one must first look to the contract to determine not only the requirement to furnish a DB delivery schedule but also to evaluate any constraints or specifications that are imposed on the form, format, and content of the schedule itself. The actual verbiage used in each clause, as well as the requirements, will vary widely, ranging from no requirement to submit a schedule to the owner in some private negotiated DB contracts, through the standard form clauses promoted by professional societies such as the Design-Build Institute of America (DBIA) and the American Institute of Architects (AIA), to highly detailed, many-page scheduling specifications in use by some state and federal agencies.

A 2001 study at the University of Texas took a comprehensive look at scheduling specifications (Ballast and Popescu 2001), subsequent to which the authors proposed a three-part specification. Table 2-1 is a synthesis of the content of the specifications in that study and furnishes a nice checklist for both owners and design-builders to use as a comprehensive baseline against which the scheduling requirements in a given DB project can be measured. The study surveyed both owners and contractors and found significant agreement in what both groups perceived to be the essential elements of a well-written scheduling specification. These are listed below, along with the clause number assigned by the researchers.

- 1.2 Scheduling Responsibility
- 1.6 Detailed Network Submission
- 2.1 Network Analysis Technique
- 2.2 CPM Software (or Equal) to Be Used
- 2.3.1 Activity Description
- 2.3.2 Activity Duration (Time Units)
- 2.3.3 Activity Coding System
- 2.3.8 Work Calendars
- 2.5.3 Detailed Network

Comparing the above list to Table 2-1 leads to the inference that a scheduling specification should be focused on furnishing a definition of the essential elements of the schedule and how they are to be related. One conclusion from the study was that "[For scheduling] specifications to be effective, they need to meet the objectives of both owners and contractors" (Ballast and Popescu 2001, PS.01.1.).

Table 2-1 Generic Contract Scheduling Clause Components

1.0 General Organization and Responsibility	**2. Scope and Products**		**3. Progress Monitoring and Updating**
1.1 Description, References, Standards	2.1 Network Analysis Technique	2.5 Network Diagram Scope	3.1 Updating Frequency
1.2 Scheduling Responsibility	2.2 CPM Software (or Equal) to be Used	2.5.1 Summary Schedule	3.2 Updating Participation
1.3 Minimum Qualifications of Planning and Scheduling Staff	2.3 Activity Related Information	2.5.2 Preliminary Network	3.3 Updated Network Approval
1.4 Training Requirement for Contractor, Subcontractor, Owner	2.3.1 Activity Description	2.5.3 Detailed Network	3.4 Updating Turnover Time
1.5 Preliminary Network Submission Deadline	2.3.2 Activity Duration (Time Units)	2.6 Project Breakdown Structure	3.5 Updating Records and Reporting
1.6 Detailed Network Submission Deadline	2.3.3 Activity Coding System	2.7 Milestones and Imposed Dates	3.6 Float Management
1.7 Review and Approval Process	2.3.4 Responsibility Codes	2.8 Activity Sorting Requirements	3.7 Change Orders
1.8 Cost of Planning/Scheduling and Monitoring	2.3.5 Activity Level Resources	2.9 Drafting Requirements	3.7.1 Change Order Representation
1.9 Progress Payments for Planning/Scheduling and Monitoring	2.3.6 Project Level Resources	2.10 Required Reports for Initial Submittal	3.7.2 Change Order Summary/ Documentation
1.10 Subcontractor Input	2.3.7 Activity Costs	2.11 Specialized Network Analysis	3.7.3 Timing of Change Order Incorporation
1.11 Contractor's Scheduling Plan	2.3.8 Work Calendars	2.11.1 Resource Aggregation	3.8 Required Reports at Each Update
1.12 Planning/Scheduling and Monitoring Audits	2.4 Required Level of Network Detail	2.11.2 Resource Leveling	
1.13 Confidentiality/ Schedule Ownership	2.4.1 Maximum Activity Duration	2.11.3 Resource Allocation Optimization	
1.14 Computer Access and Security	2.4.2 Maximum Activity Costs		
	2.4.3 Minimum Number of Activities in the Completed Network		
	2.4.4 Minimum Number of Activities in the Preliminary Network		

Source: Ballast and Popescu (2001), with permission from AACE.

Implementing Scheduling Specifications

There are two means by which scheduling specifications are typically implemented. In traditional project delivery, the method is essentially the "submit and fight" version. The constructor prepares a schedule per the contract specifications and submits it to the owner for review and approval. The owner reviews it, disapproves it, and directs the constructor to revise it in accordance with the owner's comments. If the initial schedule accurately reflected the constructor's plan when the project was bid and the owner's comments are actually owner *preferences* for the sequence of work and not about factual or mathematical errors found in the review, the constructor then resists making the directed changes until the first pay estimate is due. At that time, the payment clause allows no payment unless there is an approved schedule. So the constructor has little choice but to relent and revise the schedule to get owner approval. Now there become two schedules on the job: the "real" one that the constructor is using to control work, and the "approved" one that must be constantly revised to reflect the actual as-built schedule. This condition is not healthy for either party because if there is a problem that demands an accurate schedule to resolve, the "approved" schedule is inherently inaccurate and therefore adds confusion rather than value to the attempted resolution. Additionally, the bickering that arises over schedule approval can begin the project on an adversarial note that carries through to the end of the job.

As previously stated, DB contracting requires a higher level of professional trust than DBB project delivery to be successful. The introduction to this chapter showed that achieving schedule expectations is often the preeminent factor in project success. For these reasons, we advocate the second schedule implementation method, called "mutually developed." This approach recognizes that the schedule is critical to both the design-builder and the owner. It also accounts for the fact that, unlike in DBB, the technical scope of work is defined in performance rather than prescriptive terms and, as a result, will be clarified in the design process which must be scheduled in a manner that supports achievement of the actual construction milestones. Thus, as Gauer (2006) maintains, until it is complete and owner-approved, "the design activity is always on the critical path."

Additionally, owner design reviews will also be on the critical path. In the early stages of the project there will be a transfer of critical activities going back and forth between the design-builder and the owner, which must be carefully planned and flawlessly executed if the project is going to achieve its desired schedule. This fact alone argues for the design-builder and owner mutually developing the project's schedule rather than passing it back and forth like a minor construction submittal.

Design-Build Scheduling Conference

One way to accomplish this aim is to convene a scheduling conference shortly after the pre-work conference. The conference's primary purpose is to achieve consensus on how the schedule should be developed. Attendees from the owner's staff, the design-builder, the designer-of-record (DOR), the builder, and at times key design subconsultants, trade subcontractors, and material suppliers would meet to identify and resolve any major scheduling issues that are not addressed in the contract. An example might be the unexpected shortage of a key construction material, which must either be avoided in the design process or designed as soon as possible to allow the design-builder to lock-in the required quantity.

The second purpose of this conference is to sensitize both the owner's staff and the DOR's team about the criticality of the design reviews and how they potentially can affect project completion. Finally, the conference serves to highlight the critical points in the project's schedule for all the participants so that they can go back to their respective organizations and plan accordingly.

The agenda for the scheduling conference consists of at least three major items:

- Review and validation/revision of the conceptual schedule the design-builder used in the winning proposal
- Definition, collection, and coordination of activities from schedule participants
- Development of a scheduling framework from which the preliminary DB project schedule details can be built

Conceptual Schedule Validation

The design-builder will have developed a conceptual schedule as part of its proposal preparation activities. Often the proposal itself will contain a schedule submittal that was evaluated as part of the contract award decision. Therefore, this is the starting point for fleshing out the preliminary project schedule that will eventually evolve into the officially approved schedule. The conceptual schedule will be based on assumptions that must be validated in light of the current market situation. When the owner, designer, and builder walk through the validation process together, they will expose those assumptions that are not valid, as well as those that might have been overly conservative and can be altered to create "float" that can be used later as a time contingency for unexpected problems.

At this point, the group can also discuss options to improve the logic of the conceptual schedule. The major issue at this juncture is to ensure that the DOR and its subconsultants know which work packages must be designed early in the process to facilitate the administrative and logistics activities necessary to start the construction on time. This also cues the owner's staff as to the critical design review requirements and the time frames in which they must be completed to support timely progress. If the contract permits, this is an excellent time to discuss any over-the-shoulder design reviews that the DOR will need to ensure that the design process is meeting the owner's expectations. Finally, the members of the construction team can use this meeting to discuss coordination among the trades as well as to address those construction submittals that need prompt approval to ensure that construction progress is not delayed.

Activity Definition and Coordination

For a mutually prepared DB project schedule to truly reflect the time dimension that must be controlled during project execution, all participants must have the opportunity to contribute activities. Additionally, all participants must understand the definition of each activity to ensure a thorough understanding of the schedule itself. One technique that is used to accomplish these tasks is to develop an "activity/resource dictionary." This document is merely an inventory of each activity's detailed description along with the planned level of resources that was assumed when the activity's duration was calculated. It can also include any salient assumptions that need to be known with regard to the activity's planned completion. Table 2-2 is an example of this technique. Its major advantage is that it furnishes all the parties to the DB contract with a single reference for what activities actually denote. Additionally, it allows a valuable cross-check of the planned level of resources by more than the project scheduler, which helps validate the durations. It also allows for a forensic review of activities that are falling behind schedule to determine whether the planned levels of resources are actually being provided as well as whether the assumptions that defined the activities and their durations were accurate.

Activity Definition: Owner

The owner actually has the ability to furnish many of its activities in the DB solicitation documents in terms of project milestones, mandated owner reviews, and owner-furnished property and services. It is a good idea for the owner to take this opportunity because giving the competing design-builders a clear picture of how the owner intends to affect the project will result in more responsive proposals. That being said, if the owner has not done so in the RFP, the scheduling conference is the time for the owner to bring its expectations to light. The following is a

Table 2-2 Example Activity/Resource Dictionary

Activity Code	Activity Name (Duration in Working Days)	Description	Planned Resources	Assumptions
02A0050	Demolish Area A (4)	Includes demolition, removal, and disposal of all structural steel members in Area A as well as misc. metals found in the area.	1 Cutting/welding crew 1 5-ton end dump 50% Loader crane as req'd	Steel scrap hauled to recycler. All other materials to dump.
03B5255	Cast-in-place concrete walls Area B (2)	Includes pouring 3,000 psi PC concrete into Area B wall forms, finishing, curing, and QC testing. Also includes casting equipment pedestal B14 in same pour.	1 Concrete crew 1 Compressor with vibrator 1 Crane with concrete bucket 1 QC tech	48.5 CY of concrete; must be ordered 24 hours prior to scheduled pour.

list of common owner-controlled activities that should be incorporated into the project schedule if appropriate:

- Guidance/preferences during design to help make design decisions
- Review of design and construction work in progress
- Approval/release of final design for construction
- Required timing for design products that affect order-ship durations of owner-furnished property
- Owner acceptance inspections
- Owner close-out activities, such as commissioning
- Training of the owner's operations staff and presentation of operations and maintenance manuals
- Beneficial occupancy dates and details

The first four bullets in the above list deal with how the owner will interact during the design process. Each owner organization has its own level of comfort with giving up control over the details of design. Those that are new to DB project delivery will often impose a rigorous design review process which, if left undefined, can eat up valuable time. The purpose of this statement is not to advocate one form of control over another. Rather, it means that the owner must be forthright in declaring how it plans to administer the design in terms of the number of formal events, durations for those events, and the technical details the owner intends to accomplish in each event. This allows the design-builder to schedule its workforce to be responsive to the owner's needs during the design process.

For example, an owner of an architectural project may require the design-builder to furnish interior finish treatment color boards, renderings, and other submittals so that it can select the products it wants in the building. This is a common practice. However, the design-builder must be able to schedule the design work that must be completed before those submittals are ready for owner review, to include time to ensure that the materials selected fall within budget and schedule constraints. The design-builder also needs the owner to conduct the review and make a decision within the stated period for the review. Thus, in this example, the owner will bring the details of how it intends to make the design guidance/preference decisions and to commit to a fixed period to complete that process.

The next three bullets in the list are related to those activities that the owner will undertake at the end of the project to accept the project and make final payment. In technically simple projects these may consist of a single inspection. However, in a complex industrial process facility, such as a water treatment plant, they may start before construction is complete and extend beyond the project's physical conclusion. Many DB projects of this nature can include an operations and maintenance period after the plant is operational, to calibrate and coordinate the various pieces of capital equipment in the facility. In this case, it is extremely critical that the owner furnish the details of these types of activities and ensure that the design-builder's schedule can adequately accommodate them.

"Beneficial occupancy" is really a milestone on the DB project schedule. However, if it includes accepting various portions of a DB project before the entire project is finally accepted, it is prudent to put it on the schedule. By requiring beneficial occupancy before final completion, the owner, in essence, is mandating more than one demobilization activity after each piece of the project is turned over to the owner. This could entail a series of "pre-final inspections" and punchlist deficiency correction before the owner will accept each portion of the project. For example, consider a hospital that wants to save storage costs for the new furniture and equipment it is ordering to be installed by the owner. It therefore requires that the design-builder complete the basement by a given milestone that correlates to the arrival of the first shipment and allow the owner to store that equipment until the project is complete and it can be moved to its final locations in the building. To accommodate this requirement, the design-builder will have to ensure that all design and construction items are completed before the milestone. Additionally, doing this might require a modification to the design for the electrical and mechanical systems in the basement to separate them from other zones in the building so that the owner can store the property in a climate-controlled and powered-up area. This might also cause certain subcontractors, such those that install the interior finishes, to have to mobilize and demobilize for the basement and then come back when the rest of the building is ready. What this leads to is a requirement for coordinating the efforts of the design and construction teams in a manner that will permit a smooth flow of work both in the hospital basement as well as the rest of the project.

Activity Definition: Designer-of-Record

The DOR's major responsibility during DB project schedule development is design package definition. The construction will ultimately be completed by trade subcontractor work packages (also called bid packages). Therefore, it makes sense to schedule the design completion in a manner that directly relates to the sequence in which the project will be built. This is not to advocate that every construction work package will have its own design package, although that may be appropriate on some projects. Rather, it is to ensure that the design progresses in a manner that unequivocally supports construction progress and ensures that construction activities are not delayed because design activities are not complete. Many DB contracts are written using DBB design milestones, such as conceptual design, schematic design, and so forth for architectural projects, and 30%, 60%, 90%, and so forth for engineered projects. Both the owner and the DOR must recognize that these time-worn systems were developed with the assumption that no construction would begin until the design was 100% complete. As a result, in DBB there is no need to relate the design schedule to the construction schedule. However, as stated in the introduction to this chapter, owners typically select DB to accelerate the delivery schedule, making design and construction schedule coordination critical to timely completion. Thus, a DB contract that was formed with the traditional design milestones must be broken down into design packages to permit the necessary coordination to be effective.

It is logical that each design package will then represent an activity or a series of activities. The DOR should bring its preliminary design package plan to the scheduling conference expecting to revise it as necessary to accommodate owner or construction coordination requirements.

Another issue is the integration of the design activities with the contract requirements and the construction activities. If the contract uses the traditional design milestones, it is essential that the DOR and the owner agree on which design packages represent the achievement of each milestone. For example, a building's "30% design" submittal will typically contain a virtually complete site civil package, incomplete architectural and structural packages, and no interior finish packages. Thus, linking it to the construction activities will be meaningful only for the site civil work. To make it more cogent to the schedule, the owner could allow the 30% submittal's definition to change to represent the completion of the site civil, architectural, and structural design packages, and treat the incomplete design products that are in progress when those three packages are finished as "informational" and use them to furnish guidance or preference information regarding future design. In doing so, the DB scheduler can now directly link the construction trade work packages to the completion and approval of the 30% submittal. Now, this is obviously a bit cumbersome. A better option would be to change the contract to reflect the completion of design packages as individual milestones; this would allow measurement and payment for design progress and to do away with the traditional design milestone system altogether.

Construction Integration Activities Once the details of how the design sequence will be controlled are covered, the DOR can then drill down into the details of each design package and develop activities that can be integrated with the overall schedule. The first issue that must be covered is the impact of design on the administrative and logistics process necessary to procure materials and labor for the construction. This is where the construction team members begin talking to the design team. Two critical aspects should be dealt with before the design schedule is finalized. The first is to identify long-lead-time items of material and equipment. Typically, the design must be virtually finished on these items before they can be ordered. Therefore, it behooves the design-builder to ensure that design activities relating to these items are scheduled to be finished as quickly as technically feasible. An example might come from a highway bridge project where the main structural members will consist of precast concrete. The precast supplier will need all the detailed dimensions and other design information in order to properly fabricate the members. Depending on the volume of construction in that market, lead times for these types of members could exceed three to four months. Thus, the schedule should ensure that the structural design for the precast members be completed as soon as possible so that the design-builder could place the order and get its requirement in the precast yard's queue of work. This might necessitate approaching the bridge design in a sequence different from what the engineers are accustomed to in DBB projects.

The second critical aspect deals with features of work that require construction materials whose price is volatile in the current market. In this case, the design-builder needs to lock-in the price of those materials as quickly as possible to reduce the cost risk of those features. Again, in order to consummate a supply contract, the design-builder will need not only quantities of required materials but also specifications for its quality. Therefore, these design packages should be treated in much the same way as the long-lead-time design packages, and the DOR may have to alter the design sequence to service this need to reduce cost risk due to material cost volatility.

The next DB design scheduling activity that should be added to the schedule is one to identify and schedule the subconsultant design packages and provide time for the DOR to review, approve, and deconflict them before incorporating them into the final design. Design is by nature an iterative process, and putting these activities on the schedule recognizes that fact and adds realism to the scheduling model. One should also remember that trade subcontractors will be furnishing extensions of design in their construction submittals, and the DOR will also need to review and deconflict those. Because the trade subcontractors are under contract during design, a good practice is to accelerate the construction submittals by requiring the subs to turn them in at appropriate points in the design process. That way, issues can be resolved when there is still float in the schedule to furnish time to resolve them. Ideally, these should all be reviewed and approved before each subcontractor mobilizes to do the construction work.

Cost engineering cross-check activities The final major category of design schedule activities is to furnish time to conduct cost engineering cross-checks to ensure that the as-designed quantities of work are less than or equal to the as-bid quantities of work. The DOR's total estimated quantities of work should consist of a sum of the identified quantities of work taken from the DB solicitation and the DOR's own extensions of design, plus a design allowance for each item of work that has not been designed to the point where its material quantities can be accurately determined. The constructor must rely on the DOR to furnish the estimated quantities because the constructor is not a design professional and, as such, cannot be expected to be able to accurately gauge the potential for quantity growth during the post-award design process. Peter Ripley, an estimator who works for Fluor Canada, Ltd., (an experienced DB DOR), put it this way:

> We [the DOR] know that the quantities will grow, but we just don't know by how much. This growth is normal design development, and *an allowance is added now to include the impact of these quantities in the material takeoff, which forms the basis of the cost estimate.* (Ripley 2004, pp. CSC.08.1-CSC.08.4; emphasis added)

In fact, regardless of the constructor's ability to accurately assess the risk in its pricing contingency, the price proposal will be inaccurate if the DOR does not include a realistic design allowance for the quantity growth during normal design development. One simply cannot expect the construction professional to account for the design allowance in its pricing contingency because it requires a design professional who understands the intricacies of the given design to be able to place a value on this particular risk.

Finally, the risk that the DOR is mistaken in its computation of estimated as-designed quantities that would cause an unexpected overrun would typically be covered by the same professional liability insurance policy that would cover the risk that the design would turn out to be technically faulty because the development of the engineer's estimated quantities is a normal part of the routine design process for a transportation project in the United States. In fact, case law has found that mistakes made by the DOR in the proposal constitute design errors and omissions; this rule has been extended to inaccurate quantities (Loulakis 2006). Therefore, the need to schedule cost engineering cross-checks is critical to completing the project within budget.

These types of cost engineering cross-checks are discussed in detail in Chapter 3 and consist of the following four tasks:

- Design team members tracking as-designed quantities of work versus as-bid quantities of work as an integral part of the design process.
- Cost engineers reviewing design products on a regular basis to verify that the budget for each design package has not exceeded the as-bid budget.
- Verification with trade subcontractors and material suppliers that the materials and labor required by each design package will be available as required by the schedule.
- Identification and resolution of design issues that appear as though they will exceed the budgets and/or time frames that were assumed in the DB proposal.

Figure 2-1 illustrates the cost engineering cross-checking process. It starts as a DOR responsibility. That entity and its design subconsultants all need to be aware of the design assumptions and corresponding quantities of work that were used in the proposal. One can see the importance of this factor in Fig. 2-2, in that the lump-sum price proposal was developed based on the sum of the quantities of work developed by the owner in the RFP, the quantities from extensions of design made during proposal preparation by the DOR, and the DOR's quantity design allowance (Harbuck 2002). Therefore, the design team is expected to "design to budget" (Beard et al. 2001), and the easiest way to accomplish that charge is to ensure that the quantities of work that were used in the price proposal are not exceeded in the final design. Because the DBB design process typically waits until the final design is complete before generating the engineer's estimated quantities, many designers are not used to tracking quantity growth (a form of scope creep) during design. Therefore, putting this activity on the schedule for each design package will cue the design project manager to ensure that time is allocated to this essential task.

If the design team has checked their quantities and found them to conform to the proposal, then the estimator must next review them in light of current unit

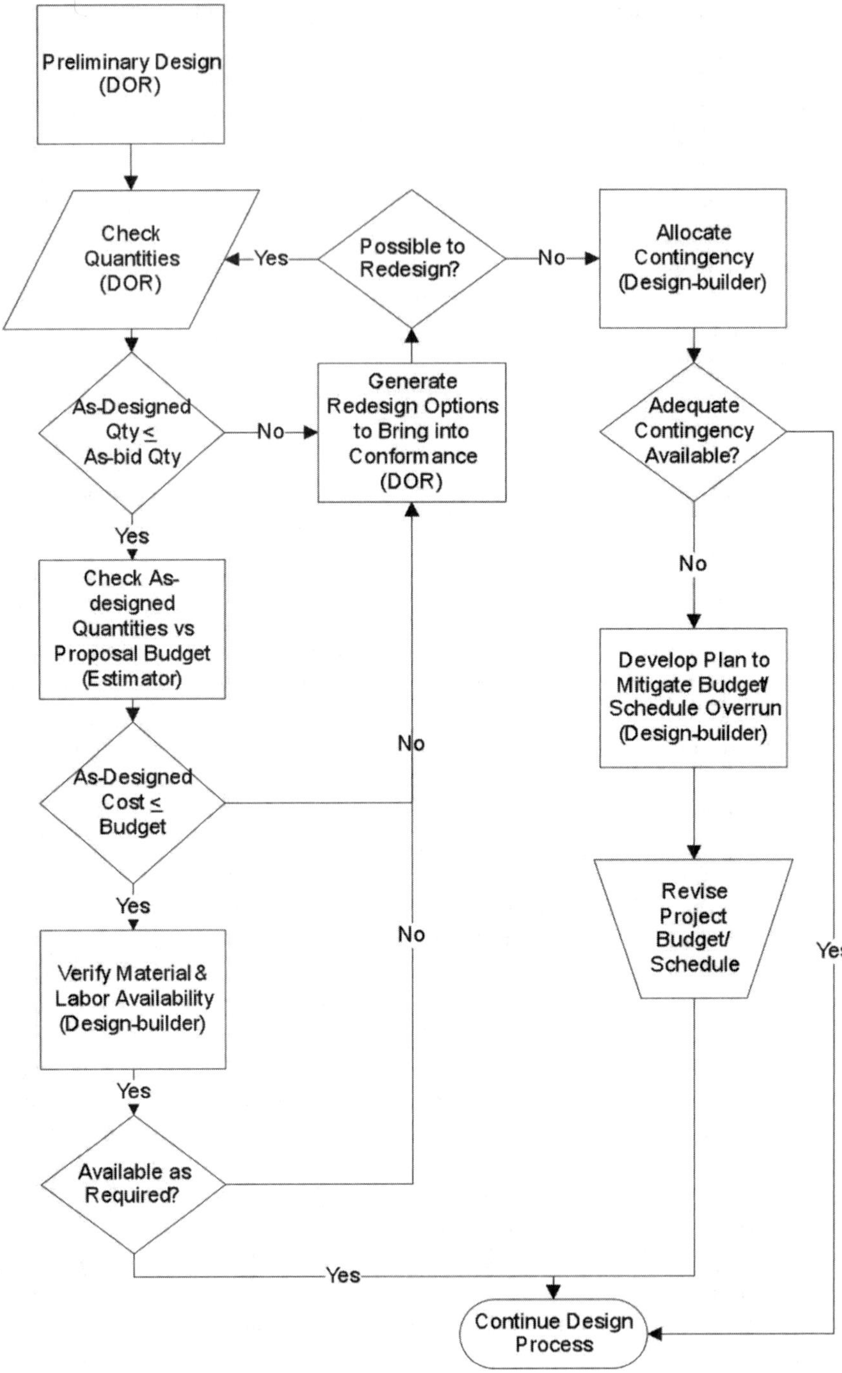

Figure 2-1 Cost engineering cross-check process during design.

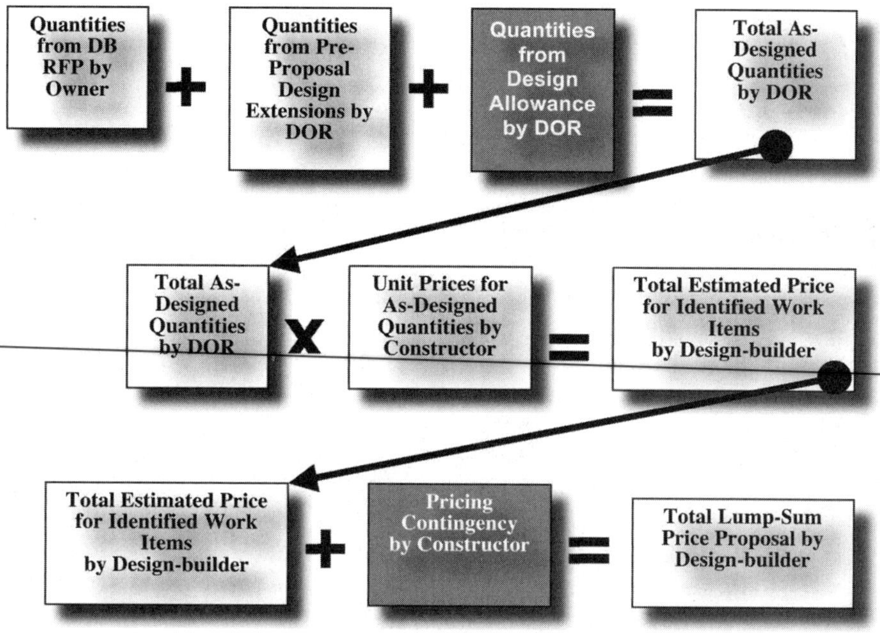

Figure 2-2 DB lump-sum price proposal development.

pricing. One of the design-builder's costs risks is volatility during design of material and in labor costs. Therefore, it is possible for a design package to hit its target quantities but still exceed the budget. In this case, the DB team needs to review the design to see whether there are ways to revise it and bring it back on budget. Finally, a check is made to see that the requisite materials, labor, and equipment can be obtained to support the project's scheduled completion date. If all the issues are resolved, the design process can proceed as originally planned or as revised to resolve budget, time, and/or resource availability issues.

Designer-of-Record Quality Management Activities DB project delivery shifts much of the responsibility for design and construction quality management (QM) to the design-builder (Gransberg et al. 2008). Often, the DOR will be assigned many of the construction quality assurance (QA) activities that were traditionally done by the owner. The DOR also must conduct the design quality control (QC) tasks necessary to ensure the quality of the design. The details of this are discussed in Chapter 4. As a result, the DOR must provide QM activities to the project schedule.

Typical design QC activities conducted by the DOR are:

- Design close-out
- Verification of critical calculations

- Review and deconflict subconsultant working drawings/specifications
- Review of guarantees/warranties
- Preparation of as-built drawings
- Review and revision of construction drawings
- Review and revision of specifications
- Design close-out

Typical construction QA activities that the DOR may need to schedule are:

- DOR construction project inspections
- Mechanical performance tests
- Testing and balancing of mechanical systems
- Electrical performance testing
- Power-up sequencing and testing
- Communications testing and calibration
- Pre-final inspection (before owner)
- Prepare DB team's internal punchlist
  - — Design items
  - — Construction items

Activity Definition: Builder

Activity definition by the builder is basically the same task as is traditionally done in DBB projects, with the exception that all construction activities will normally be preceded by some form of design or owner approval activity. There are three general categories of construction activities:

- Production activity: an activity associated with the actual construction. Requires resources and has a budget in the schedule of values.
- Logistic activity: an activity that represents the time it takes to order a required item, have it fabricated or manufactured, and shipped to the project site. Usually has no resource requirement. May have an associated budget if the costs are not included in related construction activities.
- Administrative activity: an activity that represents time allocated to complete requirements such as preparing and forwarding submittals, applying for permits or consents, or waiting for a response from an entity outside the design-builder's team.

These three activity types are easy to understand. For instance, consider the case where a design-builder must erect a structural frame for a building. The builder would start the process with the administrative activity of forwarding the steel fabricator's shop drawing to the DOR for review and approval. Upon approval, the logistic activity of ordering the steel and waiting for it to be fabricated and shipped to the site would follow. Finally, the production activity of erecting the steel would complete the process. The entire sequence would be preceded by a design-builder's design activity that details the structural steel requirement, and perhaps an owner review/approval activity for the structural design.

Other builder-contributed activities are tasks such as mobilization and demobilization where the builder must control the access to and from the job site to ensure that production is not interrupted. Additionally, the builder will typically have some public interface responsibilities, such as traffic control or relocating tenants within a building remodeling project as construction progresses. Thus, activities should be generated for these tasks and integrated with the general schedule. The builder will also have construction QC tasks that should be put on the schedule if they are critical to work progress. For instance, a requirement to pressure test an underground waterline before the trench is backfilled would directly affect the progress of that feature of work and would deserve to be put on the schedule as an activity. Additionally, the design-builder may choose to schedule joint inspections with the DOR at critical points in the project's construction, and can ensure that everything that must be prepared to accomplish each inspection's objective is ready by putting them on the schedule.

The builder might also schedule cost engineering reviews with its subcontractors and suppliers prior to key submittals to ensure that the construction budget is still valid. Finally, the builder will have a series of close-out activities that it will need to accomplish to complete the project and earn its final payment.

Schedule Integration: Partnered

Once all the DB project participants have contributed their scheduling input, the schedule itself can be developed in a truly integrated fashion. A technique that helps reinforce the required partnered nature of integrating DB scheduling is to develop the preliminary schedule in a manner that focuses on the project's contract completion date. This is done by using "backward planning" rather than the traditional "forward planning" that is used throughout the industry in construction scheduling. The theory is to start at the *end* of the schedule and begin adding predecessor activities to the final activity whose finish date is fixed as the required contract completion date. What this does is eliminate the natural inclination to be conservative when estimating activity durations. In a forward-planned schedule, the scheduler inputs all the activities and then calculates the completion date. If the calculated date is later than the required completion, the scheduler then goes back in and begins crashing activities on the critical path until the calculated date is equal to the required finish date. This procedure can often overlook activities whose durations are overly conservative that do not lie on the critical path. As the project progresses and the actual durations are added, the critical path can change and show those activities as critical when they are in fact not critical. Backward planning eliminates the post-schedule input adjustment by forcing the scheduler to question each activity's estimated duration as it is input. It also has the benefit of forcing the scheduler to adjust the network's logic as it is built by fixing that required completion date.

Ultimately if the scheduler cannot input all the project's activities without extending past the planned start date, the process shows that this project cannot be delivered on time. That realization is made at the earliest possible point so that the project participants can join forces and develop a plan to mitigate that impact. In fact, because both the owner and the design-builder are making inputs to the schedule, this will make it more credible to both parties. An example of a mitigation plan would be for the owner to agree to reduce the contractual design submittal review periods to facilitate timely completion. In a forward-planned schedule, the participants might not realize that those activities were indeed the critical ones until the project falls behind schedule after the first review.

As a part of the joint schedule development process, the owner and the design-builder need to evaluate the preliminary schedule to look for "float-makers." Float-makers are opportunities that allow activities to be completed in less than their estimated duration. For example, a DB contract that requires that the owner be allowed 20 working days to review and comment on each of four design submittals could create 20 working days of float in the project if the owner agreed to reduce its review period by 5 working days. That float could then be used by the design-builder to respond to the owner's design review comments or to accelerate the logistics process by ordering materials at an earlier date and thus reduce the uncertainty associated with the post-design procurement process. Another example would be for the design-builder to reorganize the work sequence in a manner that would allow one of its trade subcontractors to complete its work with a single mobilization instead of going back and forth to the job twice. A third typical example would be the DOR's agreeing to resequence the design process to permit the early ordering of a critical piece of equipment, thus ensuring that the equipment is on site before it is needed in case its predecessor activities are completed more quickly that estimated. This would allow the project to capture additional float that could be used to mitigate unforeseen circumstances later in the project.

The next item is to review the points in the project where the owner is on the critical path. Careful analysis will allow the owner to plan its own resource requirements and to plan to mass resources at critical points in the project schedule to facilitate timely completion. An example of "massing owner resources" would be to bring additional design reviewers from other locations or organizations to ensure that a critical design review can be completed as scheduled. If the owner needs the services of outside experts for portions of its DB QA program, knowing the points where those resources are required as soon as possible allows the owner the maximum amount of time to schedule their arrival.

Finally, the issue of "float ownership" should be discussed and settled during integrated schedule development. Generally this issue is covered in the contract and thus has effectively been decided at contract award. Nevertheless, it is helpful to discuss float ownership and ensure that all parties understand how it will operate on the given project. There are essentially four methods in use in the industry for allocating float ownership:

- Float is owned by the owner.
- Float is owned by the design-builder.
- Whoever consumes the float first, owns it.
- Float made after project award is owned by the entity that made it.

The first two methods are two sides of the same coin. In the first, the owner enjoys the advantage of float, and the opposite is true in the second. Float ownership in DBB is important when determining an equitable contract adjustment in a claims situation. Therefore, giving the owner ownership of the float allows it to mitigate the time impact of a change in scope. In the DB context where compressing the schedule is often the single most critical project performance factor, allocating the float to the owner deprives the design-builder of all flexibility. It can also be argued that it creates a disincentive to making float because the design-builder immediately loses the use of that float as soon as it is made.

The third method is commonly used if the contract is silent regarding float ownership, and it springs from case law. It essentially takes the position that, in the critical path, scheduled activities that have float are not required to start and finish as early as possible. It also recognizes that the durations used in the network are estimates, and therefore the float becomes a remedy for both parties for management errors and underestimated durations. For instance, the design-builder cannot claim a delay by an owner taking more than the contractual period to review and return a submittal if the review activity has float and the actual duration is less than the sum of the estimated duration plus the float. In the same vein, the owner cannot claim that a design-builder is behind schedule if a construction activity with float takes longer than the duration on the schedule as long as that activity is satisfactorily completed by its later finish date.

The Eastern Federal Lands Highway Division of the FHWA uses the fourth method (EFLHD 2001). This method treats float like a savings bank where the contributor gets the benefit of any float that it makes. Basically, if in 10 days the owner returns a submittal that it has 20 days to review, it can then return the next submittal in 30 days without incurring a need to change the contract completion date. The same applies to the design-builder who completes an activity earlier than its scheduled finish date. This method is an interesting approach and appears to be quite fair to both sides. It also creates an incentive toward early activity completion as a means of building "insurance" against unfavorable future events.

Regardless of the method that is used, it is essential that both the owner and the design-builder understand and accept the approach for allocating and utilizing float during project execution.

Coordinating the Schedule with the Contract Obviously, the schedule must conform to the contract requirements and assist both the owner and the design-builder in achieving the contract's requirements. Therefore, the last task in integrated schedule development is to bounce the agreed preliminary schedule against the contract itself. The first task is to ensure that the schedule can be developed at the required level of detail to allow project control functions to be effective. The rule of thumb is that the schedule's detail should be at the level at which the project will be managed. Thus, if the unit of duration is man-hours, the scheduler is saying that every worker's work can be effectively influenced for every hour of the day. This is unrealistic. So, selecting a unit of duration such as a crew-day is more logical. As the functional length of the duration increases, the schedule's level of detail decreases. This makes for fewer activities to monitor and manage. The bottom line is that both the owner and the design-builder will know what level of detail each needs to be comfortable with in their ability to control the project and understand its dynamics. It is imperative that agreement be reached on this issue so that there is only one mutually understood schedule for the contractual control and documentation of the project's progress.

Establishing Schedule Milestones Milestones are the next aspect of the schedule that needs to be covered and agreed upon. The DB contract often has contractual milestones included. To these, the participants need to add milestones that act as control points for critical actions. For example, on a budget-constrained DB project, the design-builder and owner might decide to input value engineering (VE) milestones during the design process where both entities can gather and jointly search for opportunities to add value while reducing cost. In another instance, a milestone for an owner inspection of a feature of work before it is buried can be added to ensure that the owner's QA group is aware of the progress with that feature and does not inadvertently delay the project by not being available to do the inspection shortly after completion. The decision process for establishing milestones should be controlled so that the schedule is not overly complicated as a result of adding the milestones.

To accomplish this, the parties should agree on what level of events should be considered as milestones. For instance, a design process that essentially culminates in a decision that defines a major feature of work, which then allows other design disciplines to begin their design effort on this feature, is a good candidate for a milestone; a sequence of design activities that merely results in increased design detail is not. Design decisions about items that have long procurement lead times also make good milestones due to the importance of making that decision in a timely manner. Intermediate acceptance points in construction are also good milestones. Permitting requirements can also be shown in the schedule as milestones. Finally, major sequences of activities that must be completed by various construction trades to permit a key subcontractor to mobilize and begin work should be highlighted by a milestone. Figure 2-3 is a simplified example of a schedule with milestones.

Liquidated Damages and Incentives Next, the issue of liquidated damages (LDs) should be reviewed by both parties to solidify an understanding of when and under what circumstances LDs will be assessed. This is particularly critical if the amount of LDs is onerous. The possibility for LDs can have a marked effect on the way the design-builder will approach both design and construction. It is felt especially hard during design because the construction cannot begin until the design is completed and the owner has indicated that it is satisfactory. Thus, understanding the precise

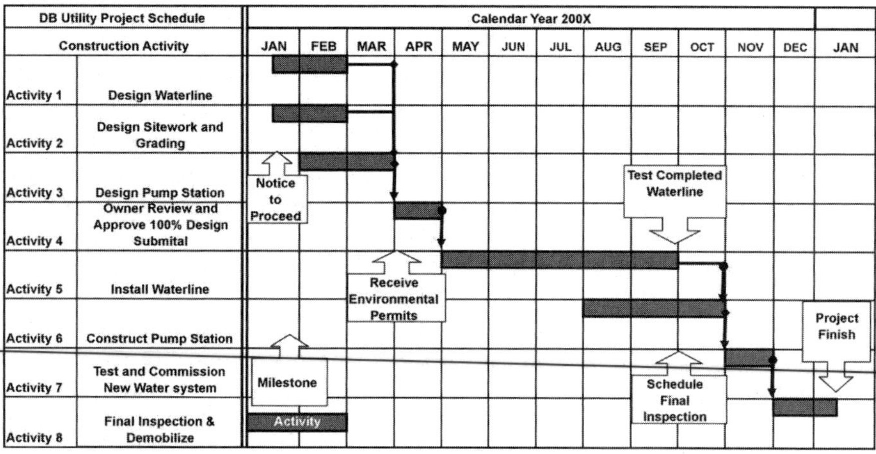

Figure 2-3 Example DB schedule with milestones.

definition of what activities must be complete to keep the LDs from being enforced is critical to scheduling the project. The federal government uses the term "substantially complete" to indicate that the project, while not totally complete, has reached a point where the owner can take "beneficial occupancy" and therefore not suffer the financial damage that LDs are designed to mitigate. So as the schedule is developed, it is critical that all the activities that lead to substantial completion are scheduled in a manner that allows the design-builder to avoid the assessment of LDs.

The next issue is the operation of any incentives for early completion that might exist. Again, the ability to enhance the project's profit margin can have a dramatic impact on the way the design-builder approaches the project. Often the presence of an early completion incentive will create justification for the design-builder to accelerate its work in a manner that allows it to capture the bonus. If the reason for selecting DB is to compress the schedule, then developing a schedule that results in early completion is certainly in the best interests of both the owner and the design-builder. This desire can be expressed at the outset of the scheduling conference and both parties can seek ways to achieve that objective.

Cost and Resource Loading "Cost-loaded schedules" are commonplace in the commercial building construction industry. The AIA standard contract between the owner and the general contractor (Clough and Sears 1994) contains a provision requiring the submission of a "Schedule of Values" against which the contractor will be paid for satisfactory progress in each pay period. To do so requires the estimator to accumulate all the direct costs associated with each feature of work that is listed in the project Schedule of Values, assign indirect costs and profit margins, and furnish a lump-sum value for each feature of work, after which the owner would pay a portion of that lump sum commensurate with the items' current completion percentage. In other words, if a work item has a value of $100,000 and is

50% complete, the owner would then pay $50,000 less any retainage that might be appropriate.

Figure 2-4 shows a DB utility project schedule as a cost-loaded bar chart. One can see that this form of display shows the expected cash flow for the life of the project. This is sometimes called "earned value" because once the constructor has completed an activity, the activity's value has been "earned," or in other words, the constructor can apply to be paid for that activity. For instance, by the end of March, this constructor would have expected to earn $165,000 because the first three activities would be complete. Cost-loaded schedules allow one to track the financial completion of the project along with its physical completion.

"Resource-loaded schedules" are typically created using commercial construction scheduling software (Marchman 1997). They can be quite arcane and complicated. The aim of resource loading is to ensure that the schedule is indeed realistic within the constraints of the available resources and their associated production rates. The purpose of this section is not to teach the reader the mathematical mechanics of resource-loaded schedules. It is, rather, to highlight the benefits that can be achieved by utilizing this powerful project control tool, and show through a simple example how this technique can be used to increase the accuracy of the estimate for an equipment-intensive project.

Resource loading has two objectives:

- To permit the accumulation of resource requirement data across the life of the project, which then permits the project manager to plan the hiring of labor and the acquisition of equipment for the project.
- To ensure that the durations for those resources that have been designated as critical are realistic for the production rates that can be reasonably achieved by the given crew.

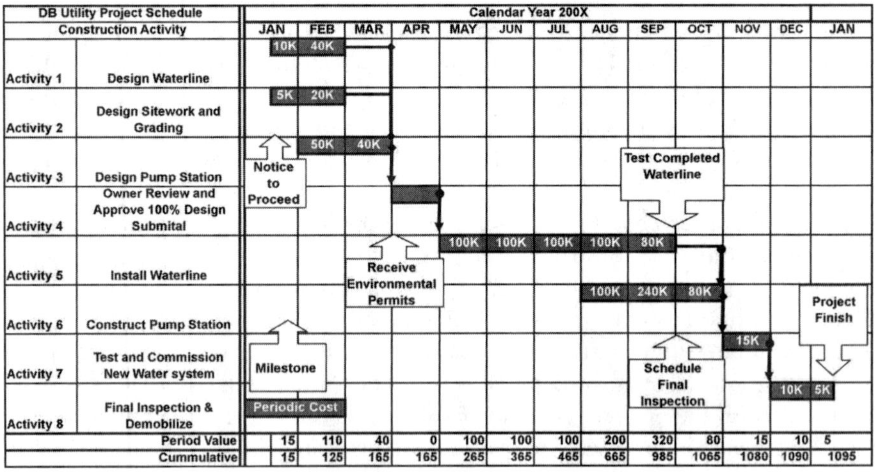

Figure 2-4 Example cost-loaded DB schedule with milestones.

Resource loading is accomplished by filling out a "resource dictionary" (Marchman 1997) in the software package that is being used by the scheduler. This task essentially consists of identifying the equipment and labor requirements needed for each crew and then associating them with the activities to which they will be assigned. The software produces a histogram for each individual trade and specific piece of equipment, which then allows the project manager to procure these resources as they are needed in the project.

Subcontractor Schedule Input The final piece of the DB scheduling puzzle is to obtain subconsultant and subcontractor input to the preliminary schedule. Once the design-builder has settled on the preliminary schedule with the owner, it needs to validate the important event dates with the design subconsultants and trade subcontractors to ensure that they can indeed comply with the constraints illustrated in the schedule. If they cannot, then the schedule will have to be revised accordingly. Additionally, the subs may be able to provide valuable potential options that were not contemplated in the schedule but that reduce schedule risk or improve the delivery period. Once the preliminary schedule is validated, the design-builder can convert it into the final baseline schedule for the project and submit it to the owner for its final review and approval.

Hierarchy of Schedules

As previously indicated, there are typically a number of versions of the overall schedule that are in use at any point in time on a DB project. Each project participant will have differing requirements for detailed scheduling. However, each participant's internal schedule or plan of work must be properly coordinated with all others to ensure that adequate resources are allocated and that no conflicts occur between members of the DB team. The foundation of all of them is often called the "baseline" or "approved" schedule.

Baseline Schedule The baseline schedule is the "official" final schedule that was developed by the design-builder and approved by the owner. It is used for classic scheduling tasks. This is the contract document that will provide the basis for computing time extensions and the delay impact in the event of a change order. This document is used to calculate earned value in conjunction with periodic progress payments. Additionally, the baseline schedule will be the reference against which the owner will determine whether the project is late and will use it to impose LDs. On a positive note, the baseline schedule will also be the benchmark to determine the amount of incentive payment for early completion. Finally, the baseline schedule will be the reference that will be used in dispute resolution or litigation to prove or disprove damages. Given all this, the baseline schedule must be clear, unambiguous, and able to be universally understood by all the project participants. This factor alone argues for keeping the baseline schedule as simple as possible. This is not to say that the design-builder's team does not need to schedule the project in detail. The idea is to furnish enough detail in the baseline schedule to permit all of the above scheduling tasks to be accomplished as required, but not so much that the schedule becomes so complex that both parties have difficulty interpreting it. The solution to this dilemma is to use the baseline schedule for contract administration and to use "look-ahead" schedules for design and construction management.

Look-Ahead Schedules The term "look-ahead" refers to the need for design and construction managers to be able to look forward into the near future to determine those tasks that must be accomplished to stay on schedule, as a part of making day-to-day project control decisions. Look-ahead schedules usually are the most detailed schedules used in the project. Figure 2-5 is an example of the four-week look-ahead schedule for the baseline DB utility project shown in Fig. 2-4. It shows the detailed schedule for the month of August. On the baseline schedule, Task 5: Install Waterline is scheduled to be ongoing and Task 6: Construct Pump Station is scheduled to begin. The look-ahead takes those two tasks and expands them in the series of activities that will take place as they are built.

Look-ahead schedules are used for routine day-to-day decision making. Their enhanced level of detail makes them an ideal project coordination tool. They are used to communicate and deconflict the design and construction work efforts. Often they will be integrated with the design-builder's project control system and permit the project manager to drill down into the project plan to identify the source of actual and potential delays in a manner that promotes timely resolution.

There are two types of typical DB look-ahead schedules: short-term and intermediate-term. Each has its specific purpose. Short-term look-ahead schedules typically display all ongoing and planned activities over a period of one to six weeks. They have the following purposes:

- Identify critical issues for resolution.
- Allocate near-term resources for the owner, the DOR, and the builder.
- Identify activities that must be expedited to remain on schedule.
- Highlight design reviews and their impact on planned construction.

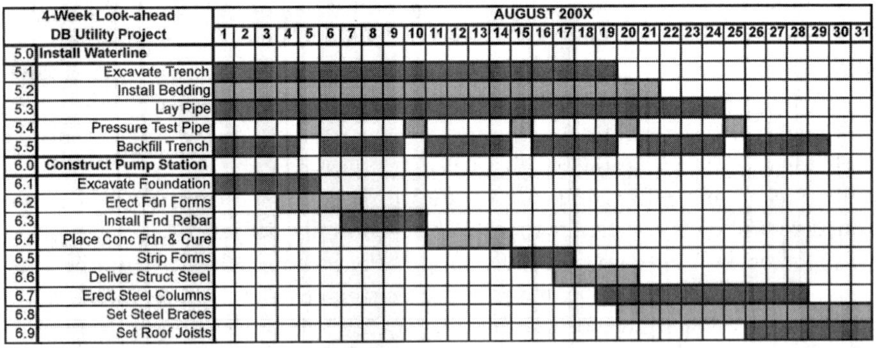

Figure 2-5 Example four-week look-ahead DB schedule.

- Schedule QA/QC and intermediate inspections for activities that are concluding in the near term.
- Highlight submittal decisions that must be obtained to maintain the planned schedule.

Intermediate-term look-ahead schedules usually cover a period of one to three months and are used for different purposes than the short-term ones. Their most important objective is to identify upcoming resource requirements so that the design-builder or owner can make the necessary arrangements to ensure that those resources are available when required. In doing so, the intermediate-term schedule is then used to manage float to ensure that the maximum level of resource availability is not exceeded.

These look-ahead schedules are also valuable tools to forecast project cash flow requirements through earned value analysis. Finally, they can be used to revise schedule milestones to reflect the impact of changes in the as-built schedule.

Schedule Updates

As the DB project progresses, it is necessary to update the schedule to account for completed activities, changes to the work sequence, changes to the scope of work, and other events that affect project completion. An updated schedule is essential to being able to evaluate the impact of scope changes in the project. It also furnishes decision makers with the project's current status and facilitates business decisions such as whether to crash activities to regain the original schedule. Finally, most DB contracts require an updated schedule as a prerequisite to periodic progress payments.

The procedure to accurately update a schedule must be well defined and followed to the letter, or the scheduler risks degrading the quality of the information provided by an updated schedule. For instance, accumulating schedule change information for a specific period (such as a month), reducing that information, and inputting it to the baseline schedule at one update adds a discipline to the update process that ensures that if the person whose intimate knowledge of the details of the schedule is unavailable for some reason, the status of the baseline schedule is still known. In other words, a policy of continuously updating a schedule as changes occur may seem logical, but it really risks a period of confusion if the scheduler leaves the project and another person must take over the scheduling responsibilities. The obvious time to conduct the regular schedule update is immediately prior to the date the periodic progress payment is due. Look-ahead schedules can be updated in draft form if the project manager needs the current schedule status to make a routine decision.

As the schedule is updated, the scheduler needs to ensure that the project's schedule of values accurately reflects the financial changes that have occurred during the previous period. This may include updating both the resource allocations as well as the activity budgets. The update should include the level of design completion, the achievement of design milestones, and a record of those design submittals that have been released for construction. It should also reflect the actual durations as start/finish dates for completed design activities. Next, construction activities are updated in the same manner as the design activities. Finally, approved change orders are added to the schedule in one of two ways, depending on the nature and timing of the change. If the change is an increase or decrease of the scope of an existing activity, the scheduler can merely adjust the duration, resource allocation, and budget for that activity to reflect its current form. While this seems intuitive, it has one drawback. The change order loses its identity, and if a forensic schedule analysis is later needed to settle a future change order to delay claim, the overall impact on the entire project of all changes is difficult or impossible to quantify. Therefore, the second method where change orders are input as new individual activities is recommended. Thus, a change that merely increases the scope and duration of an existing activity would be shown as a successor to the existing activity, with its own budget and duration.

Once the changes to the schedule have been input, the network can be recalculated and any changes to the critical path noted. It is extremely important to conscientiously perform this analysis and immediately communicate any shifts of critical activities because the entities that are responsible for those activities must revise their short- and intermediate-term plans if their activities have lost all their float due to a change in an upstream sequence of events. Once this is done, the DB project team can look at those activities that have float and determine how best to manage it. Managing float on a DB project entails comparing the early and late start dates for an activity with float and deciding what date is best to actually plan the activity's start so that the resources it needs exceed the maximum available resources. It can also be used to eliminate possible conflicts with competing activities. Finally, it can be used to mitigate the impact of noncompensable changes to the project by reassigning resources or job-site space to permit maximum production rates.

DB Schedule Imperatives

This chapter stated at the outset that most DB projects are being delivered to permit the owner to compress the project's delivery period. Thus, scheduling in DB is an essential part of achieving the owner's project goals and objectives. To ensure that the post-award execution of the project can actually attain the required completion date demands a highly coordinated effort by both the owner and the design-builder. In order to accomplish this, a set of scheduling imperatives must be satisfied:

- Joint agreement
- Joint input
- Joint understanding
- The project control tool must support the schedule
- Use the baseline as the "legal" schedule
- Use look-ahead schedules for routine project control and decision making

If the above imperatives are satisfied, both the owner's and the design-builder's needs with regard to time will be met.

References

Ballast, L. A., and Popescu, C. M. (2001). "Selecting planning and scheduling specifications." *2001 AACE Int. Trans.*, AACE International, Morgantown, W. Va.,

Beard, J. L., Loulakis, M. C., and Wundram, E. C. (2001). *Design-build: Planning through development,* McGraw-Hill, New York.

Clough, R. H., and Sears, G. A. (1994). *Construction contracting*, 6th ed., John Wiley & Sons, Inc., New York.

Federal Highway Administration (FHWA). (2004). "Design-build effectiveness study: Final Report to Congress as Required by TEA-21." *USDOT Federal Highway Administration,* <http://www.fhwa.dot.gov/reports/designbuild/designbuild0.htm> [Aug. 30, 2006].

Eastern Federal Lands Highway Division (EFLHD). (2001). *Draft design-build RFQ/RPF for Natchez Trace Parkway project PRA-NATR-3P13.* USDOT Federal Highway Administration, Sterling, Va.

Gauer, C. (2006). "Design and construction of southeast Anthony Henday Drive, Edmonton." *Proc., Transportation Association of Canada*, <http://www.tac-atc.ca/english/resourcecentre/readingroom/.../docs/.../gauer.pdf> [March 5, 2010].

Gransberg, D. D., Molenaar, K. R., and Datin, J. (2008). "Quality assurance in design-build projects." *National Cooperative Highway Research Program Synthesis 376*, Transportation Research Board of the National Academies, Washington, D.C.

Harbuck, R. H. (2002). "Using models in parametric estimating for transportation projects." *2002 AACE Trans.*, AACE International, Morgantown, W.V.

Hicks, R. G., and Epps, J. A. (2000). "Life cycle cost analysis of asphalt-rubber paving materials." *Proc., Asphalt Rubber–2000*, Rubber Pavements Association, Tempe, Ariz., 569–590.

Loulakis, M. C. (2006). "Legal aspects of design-build: What recent cases tell us." *Construction Business Owner*, <http://www.constructionbusinessowner.com/.../legal-aspects-of-design-build-what-recent-cases-tell-us.html> [March 7, 2010].

Marchman, D. A. (1997). *Construction scheduling with Primavera Project Planner®*. Delmar Publishers, Albany, N.Y.

Postma, S. E., Cisneros, R., Roberts, J., Wilkinson, R., Clevenger, J., and Eastwood, A. (2002). "I-15 corridor reconstruction project design/build evaluation 2001 annual report." *Report No. UT-02.11*, Utah Department of Transportation Research Division, Salt Lake City, Utah.

Ripley, P. W. (2004). "Contingency! Who owns and manages it?" *2004 AACE Int. Trans.*, AACE International, Morgantown, W.V.

Science Applications International Corporation (SAIC). (2003). "Outsourcing of state DOT capital program delivery functions." *Transportation Research Board of the National Academies*, <http://trb.org/publications/nchrp/nchrp_w59.pdf> [March 20, 2010].

Songer, A. D., and Molenaar, K. R. (1996). "Selecting design-build: Private and public sector owner attitudes." *ASCE J. Eng. Mgmt.*, 12(6), 47–53.

THREE

Design Administration in Design-Build

The design defines the quality of the constructed product in the design-build (DB) project. The DB project's final design is developed by interpreting the written performance criteria contained in the contract and integrating them with the design information provided by the owner in the Request for Proposal (RFP) and the design assumptions that the DB team made to prepare their proposal. This process must often be accomplished at a pace that neither designers nor owners are familiar with maintaining to achieve the aggressive delivery periods that are the hallmark of DB projects. As a result, the design administration activities take on both a level of importance and a sense of urgency that greatly exceeds their design-bid-build (DBB) counterparts. Implementing DB requires a shift in business culture to accrue the benefits of this form of project delivery, and nowhere else in the project is it more pronounced than in the design phase of the contract. Therefore, implementing a thoughtful, rigorous, and comprehensive design administration program is essential to DB project success.

In our previous book on DB contracting (Gransberg et al. 2006), project delivery was characterized as a three-legged stool, with the legs being defined as cost, schedule, and quality (as defined by the details of design) and shown in Chapter 1, Fig. 1-4. In the traditional DBB delivery system, quality is established by furnishing a completed design for the construction upon which the contractors bid (Ellis et al. 1991). Thus, with the contract completion date usually being specified, the only leg of the stool left to ensure a level platform is the bid price. Therefore DBB, by definition, is a system where the constructor tells the owner how much it will cost to deliver the quality defined in the design within the specified period of performance.

DB, on the other hand, demands that the design-builder offer a firm, fixed price for a project whose scope is defined by a set of performance criteria within a specified period of time (Molenaar and Gransberg 2001). Therefore, the variable leg in the DB stool is the details of design. This puts the design-builder in a position where the details of design, and hence the resultant level of quality, are constrained by both the budget and the schedule. In other words, the design-builder must design to cost and schedule. As a result, it is extremely important to both the owner and the design-builder that understanding and agreement be reached on the requirements for quality at the beginning of the project, so that the resultant design submittals and final construction documents will be as responsive to the owner's needs as the cost, technical, and time constraints of the project allow.

Changing the Design Administration Culture

The culture shift from DBB design administration to DB design administration requires that both the owner and the design-builder partner at the outset of the project and create an administration system that supports the development of design that is fully responsive to the DB contract. The Washington State DOT (WSDOT) DB guideline articulates this requirement as follows:

> Partnering should be considered an integral part of the Design Quality Control/Quality Assurance program. A partnering agreement is recommended to handle disputes. In addition a separate procedure for conflict resolution should be developed and agreed to by the partnering participants. (WSDOT 2001)

A national design-builder's manual for DB project administration echoes the same sentiments about the importance of partnering in DB design administration:

> It is vital for the long-term success of the project to involve the owner/client in a partnering relationship where everyone understands the roles and responsibilities of the client as well as the executing team. Indeed, design/build is the ultimate partnering relationship. . . . It is of the utmost importance to review all the technical elements of your proposal to verify and clarify any and all elements that may have been undefined, unclear or absent in the proposal. (Centennial Contractors, Inc. 2004)

Thus it can be seen that both parties to the DB contract recognize the value in formally agreeing to work together during the design phase to achieve the required levels of quality in the ultimate project. Partnering can be used as the catalyst to assist the owner and the members of the design team to make the necessary culture shift. WSDOT again provides an elegant description of the owner's role in DB design administration:

> WSDOT is expecting a proposed project that meets the design criteria and can be further developed for construction. . . . WSDOT is expecting to be available in a matter of hours or days, not days or weeks, to answer questions and provide feedback during the process. We would like to operate under a partnering environment with over-the-shoulder reviews, if possible. WSDOT will not be approving the design or construction; the Design-Builder will

> have the responsibility for ensuring the project proposal is correct. The Design-Builder will likewise have the responsibility for correcting any mistakes made in the proposal process, unless the mistakes are the result of an unclear RFP. (WSDOT 2001)

One can see from this statement of policy that this public owner is critically aware of the time element in DB design administration, and of the fact that the design-builder needs the owner to cooperate by expediting design submittal reviews whenever possible to achieve the owner's desired level of schedule compression. The scheduling issue was discussed in detail in the previous chapter, where one author was cited as stating that the design reviews were "always on the critical path" of a DB project (Gauer 2006, p. 4). Thus, it is incumbent upon the owner to implement a design administration process that supports the project's schedule objective.

Managing the Internal Design Process

The previous section described the issue of relating design administration requirements *externally* between the owner and the design-builder. The real challenge is managing the *internal* design process, that is, how the design-builder works with the designer. This topic would seem to be intuitive but, in fact, few design-builders have established formal procedures for managing the design that can be replicated from job to job. Most design-builders see this as merely managing one more subcontractor who has contractual requirements to furnish the deliverables it is assigned in a timely manner and to the level of quality specified in the general contract. This is a topic about which very little has been written, and requires formal research to develop a set of best practices that can be used throughout the industry. Without getting into great detail, here are four issues that must be addressed in the internal design administration process:

- Communications between the designer-of-record (DOR) and the owner's design reviewers.
- Developing design work packages that support the design-builder's construction plan as well as the owner's design submittal schedule.
- Integrating the builder's preconstruction review input in the design schedule.
- Making all designers on the DOR's team aware of budget and schedule constraints.

Communications with the owner during the design phase are extremely important. However, they must be conducted in a controlled manner so that the DOR's design team does not commit the design-builder to scope that is not covered in the budget. Because of the need to have free-flowing discourse between the designers and the owner's design reviewers, the best solution to keep this from getting out of control is by centralizing the communications through a design administrator that is on the design-builder's, not the DOR's, staff. The idea is for the design-builder to hire an experienced design professional to manage the internal coordination of the design and the external contact with the owner. This individual becomes the single point of contact for all design issues (often called Requests for Information, or RFIs), whether the issue comes from the owner, a design subconsultant, or a trade subcontractor. The DOR has a production responsibility that creates a conflict of interest when it comes to design quality management, in the same way that the construction general superintendent has a conflict of interest regarding construction quality management. The design administrator can act the design quality assurance manager as well as the central distribution point for communications and design deliverables. Using this approach creates a way to ensure that all communications with the owner are subjected to scrutiny by a member of the prime contract-holder's staff, and minimizes scope creep due to minor changes agreed to during design reviews.

Chapter 2 discussed the issue of breaking the design into disciplinary work packages that support the overall schedule, so there is no need to expand this here. The design administrator is an excellent facilitator for developing the design work packaging plan. In doing so, this individual can ensure that all design deliverables are included in the cost engineering and schedule review processes to ensure that they conform to budget and schedule constraints before giving them to the owner. Additionally, constructability reviews, value engineering (VE) studies, and other preconstruction services performed by the builder can be easily integrated into the design process. This coordination must be accomplished early in the process to ensure that sufficient time is allocated in the schedule for the design team to incorporate the builder's input into the final design deliverables that are furnished to the owner.

The final issue deals with ensuring that every member of the design team starts their work with a clear idea of the budget and schedule constraints. The best way to do this is to break out the as-bid quantities of work that were generated to develop the price proposal and distribute them to each design discipline. The designers are then directed to back-check for quantity growth as they proceed through the design process. If a given item of work exceeds its as-bid quantity, the designer should immediately notify the DOR and the design administrator so that action can be taken to either bring it back under the budgeted quantity or mitigate its impact on the rest of the project. Schedule constraints are best handled by identifying those features of work that contain long-lead-time items, and scheduling the design necessary to place those items on order at the outset of the design process, even if this means increasing the factor of safety to account for technical uncertainty. The other way to handle it is to implement a system of accounting for budgeted design hours for each design package and giving the design professionals periodic, preferably weekly, feedback as to how many hours remain in the budget. This allows them to keep a handle on the amount of analysis that is being conducted and lets them apply professional judgment as to the amount of time they will need to properly design the given feature.

All of the above issues are constrained by the design fee that was included in the DB price proposal. In DBB, designers tend to underestimate the amount of time they need to respond to owner's design review comments, and for construction administration after the design is complete. According to a recent study (Carr and Beyor 2005), design fees have not kept up with inflation in the same manner as construction costs. Carr and Beyor estimate that fees have dropped on the order of 20% in the last three decades. Thus, one danger that should be addressed in the price proposal is the sufficiency of the design fee. Designers may not be fully aware of the actual scope of their design administration efforts and thus may improperly account for those efforts in their fee by not considering the increased level of work required in a DB project. Many use fee curves that have been developed for DBB projects and, according to the study, may inaccurately account for inflation. The critical aspect to remember when analyzing a designer's fee schedule is that a fee curve reflects a point in time. As the effects of inflation increase price levels and salaries, the vertical scale of the fee curve must be adjusted to avoid becoming obsolete. This concept can be better understood with Carr's and Beyor's example:

> A project with a construction value of \$500,000 in 1964 had an average fee of 6.8%. In 1968 the same project was scheduled for a fee of 7.02%. This adjustment for this 4-year period was to reflect the recognition that with the increase in construction costs, a project of \$500,000 had been equal to a \$434,000 project in 1964. The fee for \$434,000 in construction in 1964 was approximately 7% ~7.15% at \$400,000 and 6.8% at \$500,000, while in 1968 a \$500,000 project was scheduled for a 7.02% fee. The fee did not change, but rather the project's construction value changed. A \$500,000 project in 1964 is not a \$500,000 project in 1968. . . . Over the same period the ENR CCI index increased from 1,581 (1971) to 3,535 (end of 1981). This would suggest that in 1971 a \$500,000 construction project would receive a professional service design fee of approximately 7.9%. The value of the work in 1981 would be estimated to be \$1,118,000 (\$500,000 × 3,535/1,581). At this value the curves would suggest a fee of approximately 6.8% for the same building. The difference is that in 1981 it cost more to build the same building than it cost in 1971, but the engineering fees were reduced from 7.9% to 6.8%. (Carr and Beyor 2005)

The above example clearly illustrates that as inflation increases the price of the construction, the fee curve must be adjusted to reflect the same inflation. Failure to adjust the vertical scale of the fee curve results in a decrease of profits for the designer (Carr and Beyor 2005). The effect that must be avoided is one where a designer has used up all the hours it allocated for its part of the design but the design is not yet complete. This results in a loss of design document quality that becomes manifest later during construction as errors, omissions, and ambiguities must be corrected at the design-builder's, not the owner's, cost.

Design Document Quality

Design quality assurance (QA) will be covered in detail in the next chapter. However, it is impossible to discuss design administration without a tying it to the ultimate objective of the design administration process: high-quality design documents.

A study was commissioned by an Australian professional society on the apparent decline in the quality of design documentation and its impact on the economy of the Australian state of Queensland. It found that "poor documentation is contributing an additional 10–15% or more to [construction] project costs in Australia" (Engineers Australia 2005). The task force that published this report identified the following root causes for the phenomenon:

1. Inadequate project briefs based on unrealistic expectations of time and cost.
2. Lack of integration along the supply chain linking the parties, and between project phases.
3. Devaluing of professional ethics and standards of business practice.
4. Lowest-bid selection strategy rather than value for money.
5. Poor understanding and skills in risk assessment and management processes.
6. Absence of an experienced, client-appointed overall Design Manager/Coordinator.
7. Poor understanding of optimized and properly documented designs.
8. Inadequate availability of, and recruitment of, skilled and experienced people.
9. Inadequate/ineffective use of technology in design and documentation.
10. Lack of appreciation of the benefits of open communication.
(Engineers Australia 2005)

This list can be used as a framework around which a DB design administration system can be defined to ensure the quality of the design documents, as the first step in guaranteeing DB project quality. In DB, the first root cause speaks to the quality of the owner's RFP and the ability of the design-builder to properly interpret it. Thus, ensuring that all the members of the owner's and design-builder's team understand the intent of the RFP is a critical design administration task. If done well, cause 7, poorly optimized and documented designs, will not be an issue. Causes 3, 5, 6, and 8 address the need for competent, well-qualified design team members. It highlights the importance of the design-builder bringing onboard the precise team that got it selected to the project. It also argues for the owner to ensure that substitutions of key personnel be carefully managed if they cannot be avoided. Cause 4, low-bid award, is a word of caution for owners about design quality. It can be rephrased like this: "If you buy a cheap design, you will get a cheap design." Cause 9 advocates maximizing the advantages of technologies such as building information modeling (BIM) to assist the design team in coordinating the efforts of various disciplines and to ensure that neither the budget nor the schedule is violated during the design phase. In Appendix A of this book, Tamera McCuen furnishes an excellent discussion of how BIM and DB can be combined to the advantage of both the owner and the design-builder. Finally, cause 10 underscores the previous discussion on partnering and demonstrates that lack of open communication during the design phase will result in poor design document quality.

In light of the above introduction, one can conclude that a DB design administration program consists of five major components:

- Qualified, competent and experienced design personnel
- Interpretation and integration of the design input from the RFP, the proposal, and the members of the DB team after contract award
- Development of the required design submittals
- Preconstruction input from the builder on each design submittal
- Review and approval/acceptance of the design by the owner

Design Personnel

Ensuring that competent design professionals are assigned to the project is the responsibility of the designer-of-record (DOR), who is the lead design professional responsible for developing the DB project's design. The following clause from a 2001 Federal Highway Administration (FHWA) contract contains a succinct definition of the DOR's roles and responsibilities:

> The Designer-of-Record (DOR) is the single point of responsibility for all design decisions and design products for the Design-Build Contractor and shall supply the required professional liability insurance. The DOR shall review, coordinate, deconflict and approve for construction all design and extensions of design produced by all members of the DB contractor's team regardless of who produces it and/or internal contractual arrangements between members of the DB contractor's team including design subconsultants, construction subcontractors, material suppliers and other entities as required. The DOR shall indicate review and approve all final construction drawings, specifications, and other appropriate design documents by fixing a stamp indicating approval for construction or the DOR's seal as appropriate. The DOR shall ensure that all design product has been priced and found to conform to the requirements of the construction schedule and the DB contractor's budget prior to submitting them to the government for review and before finally approving elements of design work for construction. (EFLHD 2001)

This clause furnishes a list of the required activities that must be controlled in the design administration process. These are "single point of responsibility for all design decisions and design products . . . review, coordinate, deconflict and approve for construction all design and extensions of design produced all members of the DB contractor's team."

The Washington State Department of Transportation (WSDOT) uses a very valuable tool to identify and allocate responsibility for various aspects of the design administration process, a WSDOT/Design-Builder Responsibility Chart. Box 3-1 shows the categories for which specific responsibilities are allocated, and Table 3-1 is an example of how it is used. This tool could either be published as part of the DB RFP to assist competitors in better understanding the owner's vision of the design

Box 3-1 WSDOT/Design-Builder Design Administration Responsibility Chart

A. Aerial Mapping

Photogrammetric Control and Panels
Aerial Photography
Plotter Compilation
— Planimetric Map
— Contour
— Topographic Map
— Drainage Area Map
— Right-of-Way Map

B. Control Surveys

Horizontal
Vertical
Topographic Map
Utility Locations
Right-of-Way
Roadway Cross Sections
Drainage Cross Sections
Structures Surveys

C. Environmental

Environmental Analysis Document
Air Quality Technical Report
Noise Analysis Technical Report
Cultural Resources Recovery
Public Meeting
— Advertisement
— Presentation Materials
— Moderator
— Technical Questions
— Transcript
— Response to Public Comments
— Liability Insurance

D. Materials Investigation

Provide Soil Survey
— Roadway
— Lateral Ditches
— Earthwork
— Retention/Detention Ponds
Bridge Foundation Investigation
Provide Testing and Analysis
Provide Pavement Design
Materials Memorandum

E. Design Traffic Data

Gather Statistics: Average Daily Traffic, etc.
Prepare Traffic Data Sheets
Prepare Equivalent 18,000-lb Axle Load

Box 3-1. WSDOT/Design-Builder Design Administration Responsibility Chart (continued)

Prepare Traffic Analysis
Level-of-Service Analysis
Composite Traffic Control Device Plan

F. **Right-of-Way**
Develop Requirements
Secure Title Search
Prepare Right-of-Way Plans and Legal Descriptions
Prepare Transfer Documents
Provide Appraisals
Negotiate Right-of-Way
Condemnation Proceedings
Testify in Court
Right-of-Way Cost Estimates
Relocation Assistance
Property Management
Clearance Letter

G. **Construction Plans**
Plot Design Survey
Basic Roadway Plans Preparation
Drainage Design
Bridge Design
Roadway Lighting Plans
Traffic Signal Plans
Signing and Pavement Marking Plans
Utility Adjustment Plans
Maintenance of Traffic Requirements
Landscape Architectural Design

H. **Environmental Permits**
Coordinate with Permitting Agencies
Prepare Permit Application
Forms
Sketches
Hydraulic Calculations
Supporting Documents
Process Permit Application

I. **Utilities and Railroad**
Utilities Identification
Submit Railroad Data
Conduct Utility Pre-Design Conference
Secure Utility Adjustment Plans
Secure Utility Relocation Schedule
Secure Utility Agreements
Process Relocation Schedule and Agreement
Clearance Letter

Box 3-1. WSDOT/Design-Builder Design Administration Responsibility Chart (continued)

J. Construction Specifications

L. Contract and Specifications

Respond to Questions on Plans, Specifications, and Estimates (PS&E)
PS&E Revisions
Addenda to PS&E, as required

M. Post-Design Services

Respond to Questions on Final Design
Review and Approve Shop Drawings
Provide Contact Person
Provide Postconstruction DB Evaluation

N. Value Analysis

Roadway Construction Plans Review
Bridge Construction Plans Review
Right-of-Way Plans Review

P. Reviews and Submittals

Roadway Construction Plans Review
Bridge Construction Plans Review
Design Concept Report Submittal
Environmental Reports
Initial Design Submittal
Preliminary Design Submittal
Final Design Submittal
As-Built Submittal

WSDOT, Washington State Department of Transportation.

Source: WSDOT (2001).

administration process, or it could be developed at the initial design conference to act as both an agenda as well as a mechanism to distribute design responsibilities to the entity that can best execute them. The major benefit of this approach is that it ensures that there are no design requirements that are not assigned at the beginning of the design administration process. In fact, the DOR could take this chart and produce one for the members of the design team, further breaking down these responsibilities among the DOR, its subconsultants, and the constructor and its trade subcontractors.

The final design personnel issue is to ensure that the key individuals who were listed in the proposal or statement of qualifications are indeed being assigned to the project. "Bait and switch" of personnel is an extremely sore topic with owners. Even given the natural personnel turnover due to promotions, retirements, job-hopping, and so forth, if a design-builder were to show up at the DB project without the named key personnel, its integrity would be called into question. If this is unavoidable, the design-builder should immediately notify the owner of the personnel change(s) and seek its approval of the substitute(s). Although there are contract clauses crafted to prevent this problem, they are hard for the owner to enforce.

Table 3-1 WSDOT/Design-Builder Responsibility Chart: Design-Build Project Example

Item	Scope Section	Design-Builder	WSDOT	Others
[. . .]				
B. Control Surveys				
Horizontal			X	
Vertical			X	
Topographic Map			X	
Utility Locations		X		
Right-of-Way			X	
Roadway Cross Sections		X		
Drainage Cross Sections		X		
Structures Surveys				X
[. . .]				
G. Construction Plans				
Plot Design Survey		X		
Basic Roadway Plans Preparation		X		
Drainage Design		X		
Bridge Design		X		
Roadway Lighting Plans		X		
Traffic Signal Plans		X		
Signing and Pavement: Marking Plans		X		
Utility Adjustment Plans		X		
Maintenance of Traffic Requirements		X		
Landscape Architectural Design				X

WSDOT, Washington State Department of Transportation.
Source: Adapted from WSDOT (2001).

The test should be to ensure that a substitute is equally qualified to a key person who is no longer on the design-builder's team. This applies to specialty design subconsultants and potentially to trade subcontractors who will supply design product to the final design.

Integrating Contract Design Input

Figure 1-2 shows the DB RFP and the winning proposal forming the technical content of the contractual scope of work. When reconciling any differences between the two, the proposal is presumed to be responsive to the RFP and, therefore, a conflict will be resolved in favor of the RFP requirement. When the design-builder assembled its proposal, a number of design assumptions were made to be able to reach a lump-sum price, but they were not required to be disclosed in the proposal. However, realizing these assumptions is critical to achieving the budget portrayed in the price proposal. Therefore, it is essential for the DOR to inventory and list the assumptions made and ensure that they are visible to all the design professionals who will contribute to the final design. An example of a typical assumption of this nature would be a parametric formula furnished to the estimator to calculate the required number of tons of reinforcing steel based on the total cubic yards of reinforced concrete. Many of the trade subcontractors will also have made certain assumptions to be able to submit their pricing. Thus it is important that the constructor canvass its subcontractors and capture these items of extended design information. Typical examples of this would be a structural steel erector who assumed that no loads would exceed the limit of its mobile crane, or a mechanical sub who used a given brand of air handling equipment on which to base its price for the capital equipment.

Once the DOR has assembled a complete set of all the design information that was developed to complete the responsive proposal, it should meet with the owner's design reviewers and ensure that the RFP performance criteria have indeed been properly interpreted, and to iron out any disputes that may arise from this exercise before commencing the design process in earnest. Taking the time to complete this task will result in several benefits to both sides. First, the DOR will be able to validate its design approach through this exercise and correct any errors it may have made in interpreting the RFP, before any design effort is wasted. Next, the owner's reviewers will have an opportunity to expand on the intent of the RFP design criteria and articulate their preferences and concerns. Finally, the outcome of such a meeting would be that both parties buy into the design approach that is associated with the contract price. It would serve to bound the technical scope of work and retard scope creep by establishing a mutually agreed benchmark against which to measure the resultant design deliverables.

Integrating Design Subconsultant Products

Next, the design subconsultants will begin to work on their individual design packages. These need to be coordinated with the larger work packages upon which the schedule is predicated to ensure that no gaps or redundancies are present. Each subconsultant must understand and agree to conform to the following procedures:

- The design quality control plan will govern the review and acceptance of subconsultant design product.
- Subconsultant design product must conform to the constraints established by the as-bid quantities of work, and the subconsultant must continuously back-check its work to achieve this goal.

- Subconsultant design product must conform to the consolidated list of design criteria and proposal design assumptions, including those furnished by the trade subcontractors.
- If either of the last two rules cannot be met, the subconsultant must immediately notify the DOR so that a meeting can be held to attempt to resolve the design issue.
- Subconsultants need to understand and use the procedures for processing RFIs and ensure that any coordination that is done with another subconsultant is recorded and passed to the DOR for its information.
- Subconsultants must adhere to the milestones established in the project's schedule and notify the DOR in a timely manner if they believe they are falling behind.
- Finally, subconsultants need to submit their requirements for design information from the owner, the DOR, or other subconsultants up-front, before the design process is started.

The DOR should also establish intermediate design hold points where all the members of the design team assemble and cross-coordinate their various products. This allows global control of conflicts, and mid-course adjustments to the overall design approach can be made to address design problems that have arisen before they become a crisis. The DOR should ensure that the constructor's preconstruction staff is on hand at these design coordination meetings. This allows them to make direct input to the coordination with regard to means and methods as well as constructability. Finally, these meetings can be used to consider potential VE ideas that have cropped up during the design process.

Integrating Design Input from the Constructors

As each design package nears its completion, time must be devoted to permit the constructors to check its cost, determine the actual availability of labor, equipment, and materials to support the schedule, and to review the constructability of the feature of work. These tasks are part of a process that has become known as "preconstruction services." This is by no means an exhaustive list of preconstruction services that can be provided by the constructors during the design phase of a DB project. These are the essential ones that are used to ensure that the project can be brought in according to budget, schedule, and construction quality constraints. Procedures must be established that create a means for the designers to incorporate the input from the constructors into the ongoing design process.

Constructability Input to the Design

The major benefit to the owner in using DB project delivery is improved constructability (Beard et al. 2001). This benefit is accrued as a result of careful design administration to ensure that all constructability input makes its way into the final design. To do so, one must understand the objective of a constructability review and its potential benefits. "Constructability" is a term of art that has come to encompass a detailed review of design drawings, specifications, and construction processes by a highly experienced construction engineer before a project is put out for bids. It is defined as "the optimum use of construction knowledge and experience in planning, design, procurement, and field operations to achieve overall project objectives" (CII 1986). The purpose of the constructability review is to cover the following five areas:

- Design errors, both material selection and dimensional
- Ambiguous specifications
- Project features that will be difficult or exceedingly costly to construct as designed
- Project features that exceed the capability of the constructor to properly build
- Project features that are difficult to interpret and will be hard for trade subcontractors (who have not yet been identified) to accurately bid

The application of constructability reviews in architectural and process plant projects has been around for decades. However, its use on transportation projects is relatively new (Gransberg et al. 1998). A comprehensive study in this field is presented in National Cooperative Highway Research Program (NCHRP) Reports 390 and 391 (Anderson and Fisher 1997a, 1997b). In this study, constructability is considered an integral part of the project development process where a project is divided into three phases: planning, design, and construction. Its authors drew this process from a study of the implementation of constructability in the non-transportation sectors. Thus, it represents not only the current thinking on this subject but also a synthesis of the latest research, making it generally applicable to any type of project.

A study by Yates and Battersby (2002) sought to identify how much construction experience a design professional should have to minimize preventable errors and conflicts during the design process as well as produce a buildable final design. The study produced an interesting set of results that bears on the implementation of constructability at the project level. One section of the survey asked a group of questions related to construction field experience, designer construction knowledge, and design processes, and reported the following results:

- 66% of the respondents thought it important for designers to have construction field experience prior to starting their design careers;
- 91% thought it important for designers to learn about construction methods, construction processes, and construction management as part of their formal education;
- 76% thought designers should be required to obtain construction field experience prior to receiving professional registration; and
- 79% felt that the number of claims against a design firm's errors and omissions insurance would be reduced if designers had construction field experience. (Yates and Battersby 2002)

This study also asked: "What is most important factor that contributes to effective construction documentation?" The most frequent response indicated that "Architects and engineers with extensive construction experience could produce the most effective documents." However, given the nature of most careers, this is difficult in the design professions where architects and engineers must spend most—if not all—of their early careers compiling sufficient professional practice experience to become licensed in their fields. The next most frequent response was to "Allow the constructor to be involved in the design from conceptualization." Thus, both of these findings would support the increased use of constructability reviews as a means of improving the quality of construction documents and achieving the benefits cited by the previous authors, and ensuring that it is part of the design administration process that is critical to DB project success.

The constructability review is incorporated into the design administration process by defining parallel constructability phases, such as constructability in conceptual design, schematic design, and construction document production. These three phases of the constructability review are further divided into features of work based on the needs of the project in question. This format can be used to guide both the constructability reviewers and the designers in ensuring that the entire project scope of work is included in the review process and that the resultant construction documents are of the best quality possible.

As the study by Yates and Battersby (2002) validated, the implementation of a constructability review cannot be conceived separately from experience in the field. Therefore, past experience and best practices are invaluable contributions to the constructability process. By incorporating information learned during similar projects, potential areas of difficulty can be identified prior to construction. Analyses and constructability reviews during all design phases only improve the quality of the final product; in this process the constructor's team tries to establish connections with similar past projects. The factors that created success in a past project can be replicated in the new project, and the reasons that led to the failure of a project can be avoided in the future.

Organizations that are new to this process can draw from the successful experience of the U.S. Army Corps of Engineers (USACE), which in the early 1980s instituted a program of conducting formal constructability reviews on all projects before they are released for bids (USACE 1994). Although no published effort has been made to capture and quantify the savings attributed to this program, our personal experience in USACE has shown that virtually every review catches some factor which, if it were left unchanged, would have necessitated a construction change order during that project.

The federal concept can easily be applied to other public and private DB projects. Essentially, it is a review of the capability of the subcontracting industry to determine whether the required levels of tools, methods, techniques, and technology are available to permit a competent and qualified trade subcontractor to build the project feature in question to the level of quality required by the contract. The constructability review also entails an evaluation of the ability of the subcontractors to understand the required level of quality and accurately estimate the cost of providing it. Thus, the level of project risk due to misinterpretation that is inherent in a set of specifications or a project feature is reduced to a minimum.

When a formal constructability review is combined with a thorough cost engineering review, the final design is greatly enhanced and the project is therefore less susceptible to cost and time growth resulting from design errors and omissions. The benefits of a constructability review, identified by Gibson et al. (1996), are:

- Reduced cost
- Shorter schedules
- Improved quality
- Enhanced safety
- Better control of risk
- Fewer change orders
- Fewer claims

A survey taken in Canada confirmed many of these benefits (Jergeas and Van der Put 2001). Additionally, their study found that

> The areas that survey respondents indicated have the greatest potential to yield the benefits are achieved by implementing the following:
>
> - Up-front (early) involvement of construction personnel;
> - Use of construction-sensitive schedules; and
> - Use of designs that facilitate construction efficiency. (Jergeas and Van der Put, 2001)

Thus, the opportunity for design-builders to utilize this powerful tool to successfully complete projects is facilitated by the DB contract structure putting the designers and the constructors on the same team. This leads to a discussion of exploiting VE ideas that arise during the constructability reviews to enhance the benefits of implementing a constructability program.

Value Engineering/Value Analysis Input to the Design

Chapter 7 contains a detailed discussion of VE in the DB context. The important issue at this point in the book is to ensure that the design administration process creates an environment where innovative alternative solutions can be expeditiously evaluated and incorporated if they add value to the final project.

The Federal Acquisition Regulation defines VE as follows:

> The systematic application of recognized techniques by a multi-disciplined team to identify the function of a product or service, establish a worth for that function, generate alternatives through the use of creative thinking, and provide the needed functions to accomplish the original purpose of the project, reliably, and at the lowest life-cycle cost without sacrificing safety, necessary quality, and environmental attributes of the project. (Blanding 2005, p. 3)

One technique to accomplish this task is the creation of a formal system for members of the team to forward potential VE ideas. The system is usually based on a form that is filled out by the initiator and passed to the DOR, who then accumulates them. At each design progress meeting, discussing these ideas is then a permanent agenda item for the group. If a given VE proposal is determined to have high potential, it is then passed to the constructor's preconstruction services group for further analysis and cost estimating, with support from the design team as required. When a VE proposal is determined to be worthy of implementation, a special meeting of the team is called to coordinate its incorporation into the design and to ensure that implementing the VE change on one feature of work does not inadvertently create an adverse impact on some other feature. The DOR must also check with the constructor to ensure that the VE change does not affect trade subcontractor scope of work; if it does, the appropriate entities are notified and given an opportunity to adjust their prices and schedules.

Trade Subcontractor Input to the Design

DB project delivery allows many submittals that typically occurred during *construction* in DBB projects to now be completed during *design* because many of the major trade subcontractors are selected before the design is complete. Both the owner and the DOR should strive to maximize the number of design decisions that are made prior to construction by integrating the efforts of the trade subs during the design process. From the owner's perspective, materials, equipment, and processes should be specifically identified as early in the design process as possible. The design-builder can write subcontracts that require early provision of submittals in a manner that allows the design team to review, approve, and incorporate these into the design submittals that the owner will eventually review.

In some DB projects the trade subcontractors furnish the design professional in their discipline. This eliminates redundancy in the design process and simplifies the integration of subcontractor input. For example, the electrical subcontractor's licensed professional engineer can do all the electrical design for the project. That way, there will be few if any electrical submittals during construction because the items that are normally submitted will already be in the construction documents. This also reduces the amount of work necessary to produce as-built drawings. Finally, it forces as many design decisions as can be reasonably addressed to be made during the design phase rather than during construction. This greatly facilitates the schedule and is extremely important in a fast-track project with parallel design and construction activities.

Design-Build Charettes

The term "charette" is derived from a French word meaning "cart" and originated in nineteenth-century France. Architecture professors at the École des Beaux Arts in Paris wheeled carts through the school grounds and collected their students' final drawings. This created a short burst of intense activity as the students hurriedly struggled to complete their work before having to put them on the cart. In twenty-first-century DB, charette has come to describe a free-flowing brainstorming session where all the project's stakeholders set aside time to participate in a facilitated, animated discussion with the objective of quickly nailing down the fundamental aspects of the design.

In DB project delivery, the use of charettes during design has several purposes. First, the charette will attempt to break down long-standing prejudices and preferences that inhibit serious consideration of potentially creative solutions to the owner's design problem. Next, by involving all the members of the DB team and the owner's stakeholders, the charette seeks to arrive at an optimum solution for the project. Finally, a DB charette provides an ideal forum for the communications that are vital to DB project success. The idea is to create an atmosphere where all concerns and ideas are seriously considered and appropriately adopted or discarded based on consensus. At the end, "buy-in" is sought from all the participants and is then reinforced during both internal and external design reviews.

Design-Build Charette Preparation

Preparation for a DB charette must necessarily start with the design-builder dissecting and distributing various sections of the DB RFP and the winning proposal to the appropriate attendees prior to the charette. This allows them time to gather their thoughts and become intimately familiar with the project's technical content. It also allows them to ensure that they have access to all necessary reference material during the charette itself. It also helps if the owner will contribute a written statement of the DB project's goals to assist the design-builder's team in understanding the critical elements of the given project from the owner itself. Finally, the charette facilitator should be given a written memorandum that lays out the design-builder's goals for the charette in terms of what it wants to accomplish by the end of the exercise. These can also be given to the attendees if it seems appropriate.

The charette goals and the owner's goals should be used as a guide to determine who exactly needs to participate in the charette. The effectiveness of many a charette has been degraded by the proverbial "cast of thousands," making it functionally impossible to achieve meaningful consensus. The group must remember that the charette is not a partnering session, although it might feel like one. Its purpose is to boil down all realistic options to the one that best fits the projects goals and constraints. The following individuals should be a part of the charette, as a minimum:

- Owner's team
  - Owner's project manager
  - Owner's design staff or preliminary design consultant who provided input to the RFP
  - Representative from the entities that will occupy, operate, and maintain the completed facility

  - Owner's construction manager or construction oversight team leader
- Design-builder's management team
  - DB project director
  - Design administrator
  - Administrative staff to document the meeting's content and to assist the participants in articulating potential solutions using computer-aided design, estimating, and scheduling software
- DOR's team
  - DOR/design project manager
  - Lead design professionals on the DOR's staff
  - Key design subconsultants for those features of work where potential design alternatives exist
  - Design quality manager
  - Graphic design technician
- Constructor's team
  - Construction project manager
  - Cost engineer
  - Scheduler
  - General superintendant
  - Construction quality manager
  - Design professionals from the trade subcontractors that will furnish input to the design, if applicable
  - Key trade subcontractors for those features of work where potential design alternatives exist
- Facilitator who preferably has experience with DB as well as running charettes of this nature

It is important that each entity has an individual present at the charette who has the authority to commit its organization to the charette outcome. If any participant has to "check with the boss" before a decision is made, it greatly reduces the effectiveness of this event. Additionally, other project participants who are not invited to the charette should be asked to be available by telephone during the charette period to permit the participants to back-check design decisions that are being made with those who will implement them. Finally, the design-builder's administrative staff must have access to the necessary electronic tools to furnish meaningful assistance to the various joint teams. This is a situation where having access to building information modeling (BIM) software is extremely valuable. BIM allows design alternatives to automatically generate cost and schedule impact information, greatly facilitating the analysis of out-of-the-box ideas which might have been rejected due to the sheer amount of time it would take to calculate their cost.

The charette should be started with the facilitator reviewing the owner's and design-builder's goals for the project. Time is taken to make sure all participants fully understand the intent of the goals; when this is achieved, the facilitator then leads the group through harmonizing the two sets into a single set of project goals that can be used to guide the remainder of the charette. The participants then divide into disciplinary teams that include personnel from both the owner's and the design-builder's teams. Each team then develops a list of important questions and needs for specific information from other teams, which are distributed by the facilitator and answered or provided by the appropriate participants. At this point, the facilitator must decide whether the charette can continue as planned or the group needs to be reassembled to settle any unanswered questions or needs for information.

The next step is to review the project's price proposal and schedule. The price is broken out into budgets for the major features of work to give charette participants a target budget within which to frame design solutions. Critical path activities are also highlighted and assigned to the appropriate disciplinary teams to ensure that they are aware of the time sensitivity that must be attached to potential design alternatives in their areas. Additionally, these teams should also look for options to expedite the design and construction of those features to create float that can be used for unexpected conditions or events while still maintaining the schedule. Finally, the cost engineer should review with the group those work packages that are associated with volatile material prices or constrained labor availability. This gives those teams the information they need to develop design solutions that can be expedited to allow the design-builder to lock-in material and labor pricing and supply.

After these opening activities have been completed, it is up to the facilitator to turn the teams loose and then gather them in periodically to review their work, guiding the group to a final design solution. The facilitator must discourage pre-charette solutions and ensure that the group seriously considers all possible alternatives. When a promising design alternative is identified, the facilitator then gathers together the requisite technical expertise, including the construction superintendent, to validate its constructability. The design alternative is then developed to a point where the cost estimator can generate a conceptual estimate to determine whether it falls within the budget established at the time of contract award. If the result is positive, that alternative then is captured for coordination with others into the final project design plan. If the result of the estimate is negative, the team decides whether its cost can be sufficiently reduced via VE.

The charette continues in this manner until the design-builder's goals for the charette have been achieved and each main entity is willing to accept the design approaches that were generated by the group. A few ground rules to assist in this important event are:

- Keep the charette format informed, focused, and well organized.
- Project director leadership is key.
- Let creative juices flow.
- Work together toward a common purpose.
- Document good ideas, hang them up, and share.

- Organize ideas into groups, such as program, site, building systems, schedule, and cost.
- Each discipline is to share criteria/constraints.
- Use time and money discussions to set boundaries.
- Conduct site analysis/orientation.
- Use structural, mechanical, electrical, and specialty consultants to set the technical constraints on the design.
- Approach the architectural design last.

Integrating Charette Input into the Design

After the charette is concluded, the design-builder's design administrator and the DOR should then review and finalize the consensus design alternatives. The result should be a set of minutes detailing the decisions and agreements made during the charette and a description of the final charette solution for each feature of work's design approach, including any conceptual estimates and schedule analysis that were conducted to validate the strongest concepts for each discipline. The package should review budget constraints for each discipline, including the target as-bid quantities that apply to each feature of work. Finally, key milestones for each feature's design should be articulated in the charette output package to ensure that the designers working on those features are fully cognizant of their time constraints. The design process is then handed over to the DOR to manage and produce the final construction documents.

Successfully integrating the design input developed in the charette involves each design professional fine-tuning the charette input to a point where it can be finalized, supplying an optimum solution. This process must be conducted iteratively between the disciplines and must include cost engineering and schedule cross-checks to keep the design feature of work within budget and in support of the contract's schedule. Additionally, the constructability must be checked to make sure that the design has not evolved beyond the capability of the construction means and methods that were assumed in the price proposal.

Given the above discussion, the charette actually serves as a road map for the design administration process by facilitating the collective design decision-making process and producing a single optimal design approach that should conform to the constraints for quality, cost, and schedule that are contained in the DB contract.

Performance Criteria Interpretation

The design criteria in the DB RFP will contain various levels of specificity. These will range from prescriptive specifications and design details that are mandated by the owner, to performance criteria that must be met by the final constructed facility and the interpretation of which is allocated to the design-builder. As previously mentioned, many—if not most—of these will have been interpreted by the design-builder during proposal preparation. Thus, there are two requirements for articulating the design-builder's interpretation of project performance criteria:

- An inventory and definition of those criteria that were defined in the proposal's design approach.
- An inventory of the remaining performance criteria that will be interpreted during the design phase.

In both cases, it is critical that all members of the design team, as well as the construction team members that will contribute elements to the design, understand the design-builder's final interpretation and apply that interpretation in their design product. This can be accomplished by assembling a "design/performance criteria dictionary" that contains the current definition for every criterion in the project scope of work.

Criteria Dictionary

This document is best maintained by the design-builder's design administrator because this individual is in the best position to coordinate the specific definitions with both the design and the construction teams. The document should be maintained in a web-accessible electronic format to make it easily referenced by design professionals, trade subcontractors, and material suppliers who need it to define the standard for which they will be held accountable. The format of this document depends on the complexity of the design requirements and the level to which the owner is relying on performance criteria to define the technical scope of work. However, it will usually include the following programmatic definitions:

- Maximum/minimum capacities
  - — Architectural space standards
  - — Structural codes: building, seismic, etc.
  - — Required clear spans
  - — Geotechnical constraints
  - — Mechanical requirements
  - — Electrical load requirements
  - — Environmental regulations
  - — Traffic flow during construction
  - — Design traffic flow
  - — Process equipment production rates
  - — Others as required
- Essential functions
  - — Access
  - — Safety
  - — Capacity
  - — Others

- Technical constraints
  - Design constraints
    - Excluded brand names and processes
    - Preapproved brand names and processes
  - Construction constraints
    - Restricted site access
    - Excluded means and methods
    - Mandatory means and methods
- Aesthetic requirements
  - Exterior finishes
  - Interior finishes
  - Landscaping
  - Art work/sculpture if applicable
- Acceptance testing criteria
  - Each design discipline
  - Major construction trades

Promulgating Performance Criteria by the Designer-of-Record

Figure 3-1 shows the hierarchy of performance elements in a DB project. The difference between a performance criterion and a performance specification is the level of specificity. Many owners and members of the DB industry misuse the term "performance specification." "Specification" implies that a design decision has been made based on the professional design process. When an owner formulates a design requirement that has more than one technically feasible solution, the proper term would be "performance criteria" rather than "performance specification" because the owner has conducted a project scoping process and determined the requirements that the design-builder will have to achieve to deliver a responsive project. The requirements of that project are then articulated in performance terms, but the owner has not completed the design process and therefore is not writing a "specification." It is important to be clear in the distinctions between the two.

The DOR will interpret the owner's performance criteria and develop performance specifications to promulgate to the trade subcontractors. For example, if the owner stipulates that the temperature within a building must be controlled, this is a criterion because it leaves open to the design-builder to choose how it will do this. If the DOR then determines that it wants to use roof-mounted zone HVAC equipment to meet the criterion, it develops a performance specification for the mechanical subcontractor, whose options are now narrowed as to the manufacturer from which it will buy the equipment. When that design decision is made, the DOR is then given a submittal to review and approve for construction. This converts the performance specification to a prescriptive specification and incorporates the approved submittal into the final construction documents. This gives both the DOR and the owner a standard against which they can inspect to ensure that the appropriate equipment was actually installed before

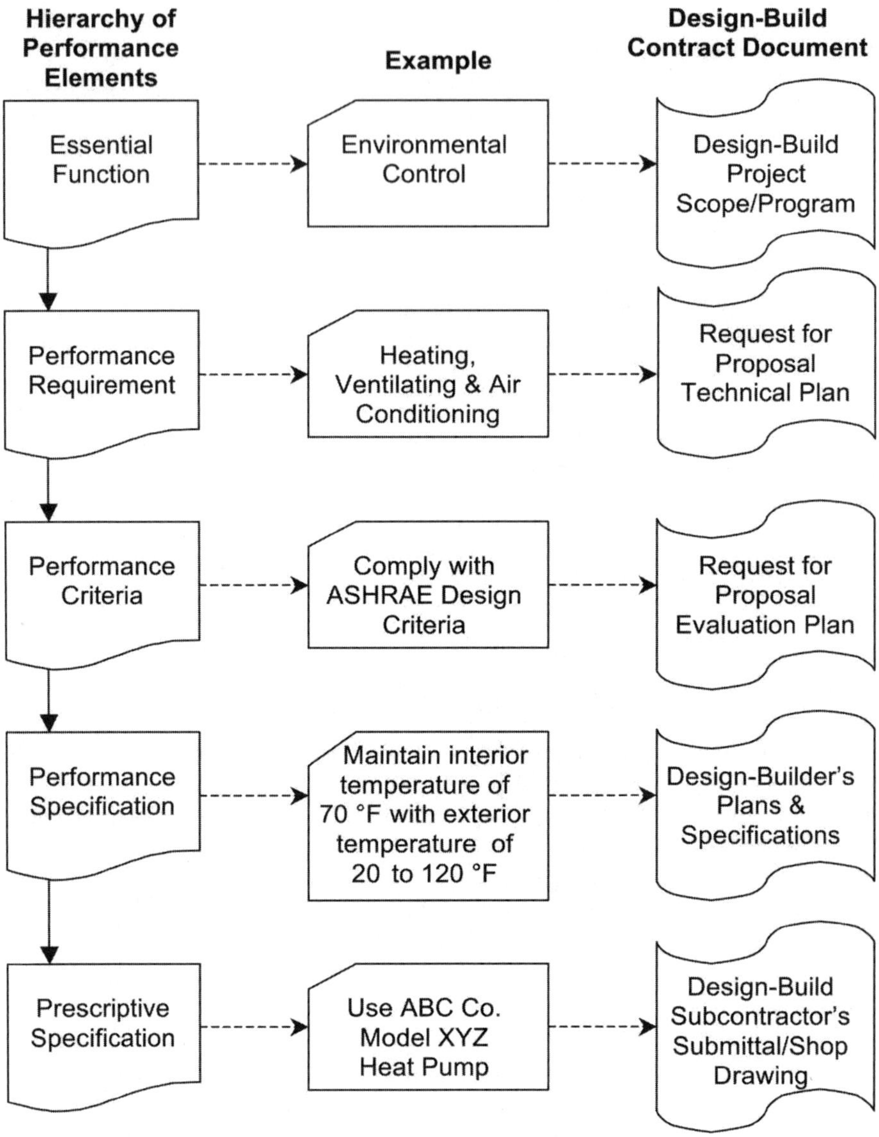

Figure 3-1. Hierarchy of performance elements in a DB project.
ASHRAE, American Society of Heating, Refrigerating and Air-Conditioning Engineers.

project close-out. Thus, the DOR controls the criteria interpretation process using the submittal process and the quality management system.

DB project delivery also furnishes an opportunity for the DOR and the trade subcontractors to generate design alternatives that meet RFP performance criteria as well as budget and schedule constraints, and present the owner with viable options and receive owner guidance as to its preferences. In these cases, all the alternatives are of roughly equal value to the design-builder, and the design-builder seeks to strengthen its relationship with the owner by implementing owner preferences wherever possible. Often these preferences were expressed during one of the initial conferences and were put on hold until the design had progressed to a point where they could be addressed. The idea here is that the owner does not get a guarantee that its preferences will be implemented, but is assured that its preferences will be considered and, wherever possible, incorporated into the final project.

Design Submittals and Reviews

The design submittal process is discussed in detail in Chapter 4, Design Quality Management in the Design-Build Project. These deliverables are the heart and soul of the design administration process. Chapter 2, Scheduling Design-Build Projects, also discusses the fact that design deliverables are always on the critical path and hence constitute important milestones on the schedule. There are many technical issues with design submittals as well. Obviously, they must be compliant with the contract's design and performance criteria; that is a quality management issue addressed in the next chapter. The time it takes to produce the design deliverables and then revise them in accordance with the owner's design review comments must also fall within the project's budget and schedule. Finally, the design team must be aware of the impact of the design product on the construction process, ensuring that their product can indeed be constructed using the means and methods that were assumed during the preparation of the DB proposal.

Design-Builder's Perspective on Design Submittals

The design-builder is primarily concerned about each design submittal's impact on the project's schedule and budget. Therefore, it is incumbent upon the design-builder to establish an internal design administration system that ensures that no design product leaves the team without passing through the constructor's cost engineer and scheduler. To achieve this requires a strong design administrator. This person must establish and enforce clear lines of communication and protocol between the design team and the construction team. Doing so ensures that both the constructor's and the owner's project management group are included in the design submittal loop. The first duty is to ensure that the design complies with the design-builder's budget and time constraints. If this is not possible, the design administrator will then convene a meeting with the DOR and the constructor's staff to identify possible options or to mitigate the impact of the overrun.

The design administrator will also closely monitor the progress of trade subcontractor design input and construction submittals to ensure that they are in compliance with the contract and the proposal. The DOR expects these vendor documents to supplement their drawings and specifications. Therefore, as many of these construction-type submittals as possible should be prepared, developed, reviewed, and approved during the design phase to minimize the potential for delays due to problem submittals after construction has started.

Finally, both the owner and the design-builder's team need to remember that during the design phase, working drawings must be allowed to be revised as the design progresses. There are a number of reasons why a working drawing in an early design submittal will need revision, chief of which is the iterative nature of the design process itself. Additionally, the need to maintain both budget and schedule may drive the decision to revise an early drawing. The challenge with revisions is making them without creating a loss of trust. Applying the principle of design commitment uniformly throughout the design administration process solves this problem. The owner's design reviewers know that if commitment has not been made to a given feature of work, they can expect that feature to be revised as the design is further developed. On the other hand, if a committed design must be revised, the design-builder must justify the revision to the owner and get its concurrence. To make a design commitment, the design administrator must ensure that the design has been priced and reviewed for compliance with the current design and performance criteria. The availability of materials, equipment, and labor should also be verified to ensure that committing to the design will not adversely affect the project schedule.

Owner's Perspective on Design Submittals

The owner's design reviewers do not conduct a technical review; they conduct a "compliance review." Therefore, the major issue for owners to ensure that the design under review meets RFP design and performance criteria. Additionally, it should reflect any betterments that were offered in the design-builder's proposal. Owners must be careful to not unintentionally assume design liability by their review comments. Additionally, the reviewers must remember that a DB contract is scope-driven and therefore they must maintain an open mind to accepting design approaches with which they may have no personal experience. They should expect a somewhat different design product. In DB the DOR does not need to produce the "biddable" construction documents required in DBB projects. Therefore, the owner's reviewers should expect to see less detail in the drawings, and specifications that consist of a combination of everything from traditional performance specifications to manufacturers' catalog cuts for material and equipment that has been furnished by trade subcontractors to the design.

The purpose of the owner's design review is to determine whether the design will ultimately result in a satisfactory constructed facility. Therefore, its reviewers should view design documents from the standpoint of making sure they contain enough technical detail to allow the constructors to properly build the project. The reviewers must also keep in mind that the design team is working for the construction team, and the designer–owner client relationship does not exist in a DB project. The advantage here is that the design team can more actively participate in the construction and even decide to put off certain elements of design until construction is complete enough to allow the design to be finished using actual field dimensioning if that makes sound engineering sense. The disadvantage is that the DOR is no longer the owner's advocate. This is why some opponents of DB charge that "the fox is guarding the henhouse." Research has shown that this apprehension is unfounded and that DB project quality has been found to be roughly equal to the quality found on the DBB projects. The FHWA *Design-Build Effectiveness Study* reports actual results that conclusively confirm this belief, as summarized in:

> On average, the managers of design-build projects surveyed in the study estimated that design-build project delivery reduced the overall duration of their projects by 14 percent, reduced the total cost of the projects by 3 percent, and *maintained the same level of quality as compared to design-bid-build project delivery.* (FHWA 2006; emphasis added)

Once again, it is important for the owner's reviewers to approach the review with an open mind and not view changes to previous design submittals as attempts to enhance profits by degrading quality. As stated in the introduction to this chapter, the variable leg in the DB project "stool" is the details of the design. By implementing the design commitment principle, the owner can use it as the litmus test for judging revisions to previously submitted designs. This concept helps preserve the atmosphere of trust and cooperation that is so necessary in a fast-moving DB project.

Design Progress Review Meetings

If the formal design submittal is the heart and soul of the design administration process, the periodic design review meeting is its bones and flesh. The purpose of periodic design review meetings is to efficiently coordinate the details of the design. They also serve as an excellent forum to float ideas to reduce costs, save time, or add value to the final project. To maximize the effectiveness of these meetings, all the right project participants need to be present. A typical list of attendees is:

- Project director
- Design administrator
- Lead for each design discipline
- Constructor's preconstruction manager
- Subcontractors who will contribute to the design (if appropriate)
- Specialty subconsultants and/or subcontractors (if required)
- Owner's representatives (when scheduled)

The agendas for these routine periodic meetings should be standardized to ensure that each meeting covers all the necessary points of information to allow the designers to achieve maximum production in the period following the meeting.

Here are typical agenda items that could be included in a standard design review meeting:

- Permit schedule
- Design milestone schedule
- Logistics milestone schedule
- Action item lists
- Status, questions, and concerns by discipline

These meetings are usually held on a weekly or biweekly basis, depending on the pace at which the design must be advanced. They are usually scheduled to take about one hour for the entire group. Some design-builders will then plan to use the next hour for ad hoc interdisciplinary meetings to iron out coordination details that do not require the entire group's attention. If a particular problem is identified in this meeting, then follow-up meetings are scheduled with the appropriate members of the design and/or construction team. As the design approaches a milestone, additional meetings are scheduled to bring together the necessary disciplines to accomplish the following tasks prior to submitting the design deliverable for the owner's review:

- Drawing reviews
- Specification reviews
- Coordination reviews

Fast-Track Design Reviews

When an owner decides to compress the project delivery period to as short as possible, it must recognize the need to field a design administration system that will promote rather than hinder design progress. Construction cannot start until the necessary elements of design are completed, reviewed, and released for construction. Chapter 2 discussed the schedule implications of fast-track projects, so there is no need to repeat them here. Essentially, the owner and the design-builder should mutually answer the following questions:

- What must be formally reviewed?
- What features of work require owner review and approval?
- On a feature-by-feature basis, can construction begin before the review is complete?
- What critical logistic activities are tied to design review completion?

The first question seeks to reduce the number of design review actions to the minimum requirement. Some fast-track projects have actually done away with formal design reviews and substituted co-located over-the-shoulder design reviewers. In this approach, the owner and the DOR literally set up offices in the same location. Often both parties' personnel are actually in the same office with adjoining work stations. This allows for questions to be asked as they arise and for answers and design decisions to be made and accepted on a continuing basis. One of this book's authors used this approach to deliver an emergency replacement of a community's water treatment plant in less than 12 months. This begs the question: If this approach can be successfully used in an emergency, why not use it in non-emergency projects with aggressive delivery schedules? Obviously, to co-locate the owner's and DOR's personnel will add a marginal cost to the project. Therefore, a cost-benefit analysis should be done to determine whether the time saved justifies the additional project overhead cost.

The second question asks the owner to take a hard look at the various features of work included in the project and identify those features where the owner is not comfortable shifting the entire design responsibility to the design-builder—trusting the contract and the design quality management system to provide the required level of design and construction quality. Often public works projects have the owner's standard design details incorporated by reference in the contract. Additionally, most projects include large amounts of what could be called "standard design requirements" that promulgate building and life safety codes, and greatly limit the ability of the designer to use an approach that has not previously been found to be acceptable. Thus, under the second question, the owner should really decide where the areas of greatest design risk are and focus its attention and resources on those features. It can then allow the design-builder to assign competent, experienced design professionals to the routine features of work and proceed without interference by the owner.

The third question should be answered at the initial design meeting on a fast-track project if it has not been answered in the RFP's scheduling section. Permitting requirements will often drive the answer to this question. Thus, the DOR needs to start with a list of those areas that must be designed first to permit their design to be released for construction at the earliest possible opportunity. The answer to this and the final question will ultimately drive much of the fast-track project's schedule. The need to define the requirements for materials and capital equipment so they can ordered, fabricated, and shipped in time to permit their timely installation will also create a need for early design completion on those features of work that have these items in their scope.

To summarize, both the owner and the design-builder must adjust and fine-tune the design administration system to support the schedule when attempting to complete a DB project on a fast-track basis.

Drawing Numbering System

As has been said before, DB project delivery requires a higher level of trust between the parties to the contract than in other project delivery methods. The owner is giving up control over the details of design, and this creates a situation where the owner must be able to implicitly trust the design-builder to exercise the same professional care and diligence that the owner would expect in the design contract of a traditional DBB project. The "fox guarding the henhouse" fear often makes the owner distrustful of changes to an aspect of the design after it has reviewed and accepted it. Again, the use of design commitment will help to allay this fear.

Additionally, there is a need to distinguish between those design products for which commitment has been made and those that are still a work-in-progress. One simple technique is to use the drawing numbering system to form that identification. In this method, anyone can pick up a drawing that has been left behind in a conference room and tell by looking at the number whether the details on that sheet are subject to change without notice.

Drawing Types and Sources

Four types of drawings will typically be generated in a DB project:

- Conceptual design drawings
- Working detail drawings
- Final construction drawings
- As-built drawings

They will be created by a number of sources. The owner will have included a number of drawings in the project solicitation documents that will become part of the contract. Additionally, it will often furnish standard details and drawings for equipment for which it wants to ensure compatibility. Obviously, the DOR and its subconsultants will be producing a full range of drawings during the design phase. Trade subcontractors may also be furnishing elements to the final design as well as technical submittals that are necessary to complete the design. This results in a probability that someone may see a drawing, believe it to be complete, and then be disconcerted when the final construction documents are reviewed and that particular detail has been changed. This threatens the mutual trust between the two parties.

Therefore, the objective of the drawing numbering system is that its methodology be such that anyone can look at a drawing and determine whether or not it has been priced (committed). This is essential to maintain the required level of trust. It does so by enhancing the communication of technical detail. It also reinforces the chain of authority by making the DOR the keeper of the drawing numbers and, hence, the authority for design commitment. Further, it forces the DOR to take and maintain control of the complete design because the DOR must assign the drawing numbers that correspond to the level of design completion.

Drawing Numbering System Concept

Many public and private owners have standardized drawing numbering systems that have been used for decades. There is no need to change these just for a DB project. In fact, the most effective system is one where the number of a drawing that is not part of a committed design looks visibly different from what the owner is accustomed to seeing. This can be done by merely adding a prefix to the drawing number until it has been committed, and then removing the prefix when design commitment is made and it is prepared for the owner's review and release for construction. For example, the following list is for a structural drawing as it proceeds from concept to as-built:

- Concept drawing: C-S201
- Working drawing: W-S201
- Record drawing: S-201
- As-built drawing: S-201 with redlines

In addition to the drawing number, the presence of the appropriate design professional's seal will reinforce the fact that, for the time being, the DOR believes this drawing to be the one that will ultimately become a part of the final construction documents. Some owners and design-builders require an area be made available in the title block for the DOR to sign, indicating that it has reviewed the drawing and found it to be in compliance with the contract requirements as well as ready to be released for construction.

In summary, the DB project's design defines the quality for the DB project's construction. Most DB projects that are awarded have challenging delivery schedules and, as a result, completing the design in an expeditious manner is essential to the project's success. Design administration is vital to achieving project success, and has three levels. First is the owner–design-builder interface where design information and product must be exchanged in precise terms. Next is the design-builder–DOR interface where the design-builder must ensure that the work controlled by the DOR meets the project budget and schedule constraints as well as complies with the design and performance criteria in the DB contract. Finally, there is the DOR–constructor interface where the DOR leverages the construction knowledge and experience of the constructor and its subcontractors to enhance project constructability and reduce risk during construction. When all three levels are working in harmony, a win-win situation is achieved. The owner gets its facility on time, within budget, and to the standard of quality prescribed in the contract. The design-builder and its team achieve their target profit for the work and enhance their reputation with this owner, making them more competitive for the owner's next project.

In the final analysis, the operative term is "project quality" because it is the quality of the constructed product that endures for the project's lifetime of operation and maintenance. Therefore, the next chapter will discuss design quality management to show how design quality can eventually be turned into construction quality.

References

Anderson, S. D., and Fisher, D. J. (1997a). "Constructability review process of transportation facilities." *National Cooperative Highway Research Program Report 390*, Transportation Research Board of the National Academies, Washington, D.C.

Anderson, S. D., and Fisher, D. J. (1997b). "Constructability review process of transportation facilities workbook." *National Cooperative Highway Research Program Report 391*, Transportation Research Board of the National Academies, Washington, D.C.

Beard, J. L., Loulakis, M. C., and Wundram, E. C. (2001). *Design build planning through development*, McGraw-Hill, New York.

Blanding, J. R. (2005). "Using VE in design build." *Proc., 2005 AASHTO Value Engineering Conference, San Antonio, Texas*, American Association of State Highway and Transportation Officials, Washington, D.C.

Carr, P. G., and Beyor, P. S. (2005). "Design fees, the state of the profession, and a time for corrective action." *ASCE J. Mgmt. Eng.*, 21(3), 110–117.

Centennial Contractors, Inc. (2004). *Design-build project manual*, Centennial Contractors, Inc., Vienna, Va.

Construction Industry Institute (CII). (1986). *Constructability: A primer*, Construction Industry Institute, Austin, Tex.

Eastern Federal Lands Highway Division (EFLHD). (2001). "Draft design-build RFQ/RPF for Natchez Trace Parkway project." *PRA-NATR-3P13*, USDOT Federal Highway Administration, Sterling, Va.

Ellis, R. D., Herbsman, Z., and Kumar, A. (1991). " Evaluation of the FDOT design/build program." *Final Report, submitted to Florida Dept. of Transportation, State Project No. 99700-7543-010*, Department of Civil Engineering, University of Florida, Gainesville, Fla.

Engineers Australia. (2005). "Getting it right the first time." *Construction document quality report*, Queensland Division Task Force, Brisbane, Queensland, Australia.

Federal Highway Administration (FHWA). (2006). "Design-build effectiveness study—As required by TEA-21 Section 1307(f): Final report." *USDOT Federal Highway Administration*, <http://www.fhwa.dot.gov/reports/designbuild/designbuild0.htm> [Aug. 30, 2006].

Gauer, C. (2006). "Design and construction of southeast Anthony Henday Drive, Edmonton." *Transportation Association of Canada*, <http://www.tac-atc.ca/english/resourcecentre/readingroom/.../docs/.../gauer.pdf> [March 5, 2010].

Gibson, G. E., McGinnis, C. I., Flanigan, W. S., and Wood, J. E. (1996). "Constructability in the public sector." *ASCE J. Const. Eng. Mgmt.*, 122(3), 274–280.

Gransberg, D. D., Senadheera, S. P., and Karaca, I. (1998). "Analysis of statewide seal coat constructability review." *Texas Department of Transportation Report No. TX-98/0-1787-1R*, Texas Department of Transportation, Austin, Tex.

Gransberg, D. D, Koch, J. E., and Molenaar, K. R. (2006). *Preparing for design-build projects: A primer for owners, engineers, and contractors*, ASCE Press, Reston, Va.

Jergeas, G., and Van der Put, J. (2001). "Benefits of constructability on construction projects." *ASCE J. Const. Eng. Mgmt.*, 127(4), 122.

Molenaar, K. R., and Gransberg, D. D. (2001). "Design-builder selection for small highway projects." *ASCE J. Mgmt. Eng.*, 17(4), 214–223.

U.S. Army Corps of Engineers (USACE). (1994). *Engineer Regulation 415-1-11: Biddability, constructability, operability, and environmental review*, Department of the Army, Washington, D.C.

Washington State Department of Transportation (WSDOT). (2001). *Guidebook for design-build highway project development*, Washington State Department of Transportation, Olympia, Wash.

Yates, J. K., and Battersby, L. C. (2002). "Master builder project delivery system and designer construction knowledge." *ASCE J. Const. Eng. Mgmt.*, 129(6), 145–151.

FOUR

Design Quality Management in the Design-Build Project

Design-build (DB) project delivery alters the traditional roles seen in design and construction quality management. By selecting DB, the owner shifts the design from its own team to the design-builder's team. Thus, this requires a shift in the way quality management is performed throughout the life cycle of a DB project. In the traditional design-bid-build (DBB) system, the owner is responsible for design and construction quality assurance (QA). The Designer-of-Record (DOR) is responsible for design quality control (QC) and often assists the owner with design QA, which leaves the constructor responsible for construction QC. Implementing DB entails shifting quality management tasks. One of the major motivations for using DB project delivery is to shift design liability from the owner to the design-builder.

To effectively transfer design liability to the design-builder, an owner must also transfer many of the traditional quality management responsibilities as well. This leads to a concern that the "fox may be guarding the hen house"; this concern has plagued DB since its inception. A study by Ernzen and Feeney (2002) of the Arizona DOT's DB program (appropriately titled "Contractor-Led Quality Control and Quality Assurance Plus Design-Build: Who Is Watching the Quality?") addressed this concern directly by comparing project QA test data on a DB project where the design-builder had been assigned the responsibility for QA, with data from a similar project delivered by traditional DBB means. It found the following:

> Analysis of the data shows that despite a highly compressed schedule, the quality of the material on the project exceeded the project specifications and was similar to the quality of work completed for the state under traditional contracting methods with an Arizona DOT-operated quality assurance program. (Ernzen and Feeny 2002)

The Arizona DOT study and the numbers regarding the growth in DB across the nation, previously cited in Chapter 1, effectively belie the theory that implementing

DB project delivery will inherently result in poorer construction quality. It would be difficult to believe that sophisticated public and private owners would propagate the spread of a delivery method that consistently resulted in substandard or poor-quality product regardless of its ability to expedite project delivery. In fact, a Federal Highway Administration (FHWA) study on DB effectiveness reported actual results that conclusively confirm this belief, as summarized in the following quotation:

> On average, the managers of design-build projects surveyed in the study estimated that design-build project delivery reduced the overall duration of their projects by 14 percent, reduced the total cost of the projects by 3 percent, and *maintained the same level of quality as compared to design-bid-build project delivery.* (FHWA 2006; emphasis added)

Design-Build Quality Assurance Model

Previous authors have described the change from DBB to DB procurement as a "culture shift." Our 2006 companion book, *Preparing for Design-Building Contracting: A Primer for Owners, Engineers, and Contractors,* sums this up as follows:

> . . . [D]esign-build contracting deals with the ability to the parties in the contract to execute it in an environment of trust. This is the ultimate culture shift that must be made by owners, designers, and builders. (Gransberg et al. 2006)

Concern for project quality is the factor that causes distrust. To allay this fear, DB projects must be delivered with at least the same level of quality that occurs in DBB. In traditional DBB contracting in the construction industry, decades of QA and QC experience provide a wealth of knowledge and standard practices that are readily accessible and widely accepted for assuring project quality. In DB, on the other hand, there exists a limited but rapidly expanding body of experience associated with assuring quality. One of the major challenges facing owners and design-builders in implementing DB is the change that must take place in the culture of both parties. This is described well in an evaluation report given by Postma et al. on the I-15 DB transportation project in Utah:

> The Owner felt that one of the biggest challenges to the QC and QA program was "breaking the mold" of the traditional roles of the contractor and Owner. The Owner's personnel had all come from the "catch and punish" culture. Likewise the Contractor personnel came from a similar background. To change philosophies to a more proactive quality role by the Contractor and a less controlling oversight role of the Owner was a significant challenge. Most personnel assigned to the project by either party had worked under

traditional systems for many years and this was the first experience with this type of project. (Postma et al. 2002)

As both owners and design-builders become more familiar with DB, the culture change will be less of an issue.

Another structural issue that has confronted DB projects since their implementation has been the idea that construction contractors, who lead many DB projects, would pressure the designers, as their subcontractors, into sacrificing quality for higher profits. According to the previously quoted FHWA *Design-Build Effectiveness* study (FHWA 2006), this has not been the case. However, a well-defined quality management system is one way to address these fears. This is even more important on high-profile public projects delivered using DB. According to Leon Pantazides, a quality management system "adds credibility and assurance for all involved from the design-builders, to the department of transportation [i.e., the public owner], to the public users" (Pantazides 2005).

Design-Build Quality Definitions

The *Transportation Research Circular E-C074: Glossary of Highway Quality Assurance Terms* (TRB 2005) contains standard definitions for quality assurance. Other organizations, such as the American Society for Quality, also publish their own definitions. Additionally, many public owners such as the U.S. Army Corps of Engineers (USACE) utilize their own definitions for major quality terms. For purposes of this book's discussion of DB quality, the *Transportation Research Circular E-C074* definitions will be used to standardize the information presented on this subject. As such, the major quality definitions are as follows:

- *Quality.* (1) The degree of excellence of a product or service; (2) the degree to which a product or service satisfies the needs of a specific customer; or (3) the degree to which a product or service conforms with a given requirement.
- *Quality Assurance* (*QA*). All those planned and systematic actions necessary to provide confidence that a product or facility will perform satisfactorily in service. (QA addresses the overall problem of obtaining the quality of a service, product, or facility in the most efficient, economical, and satisfactory manner possible. Within this broad context, QA involves continued evaluation of the activities of planning, design, development of plans and specifications, advertising and awarding of contracts, construction, and maintenance, and the interactions of these activities.)
- *Quality Control* (*QC*). Also called *Process Control.* Those QA actions and considerations necessary to assess and adjust production and construction processes so as to control the level of quality being produced in the end product.
- *Independent Assurance* (*IA*). A management tool that requires a third party, not directly responsible for process control or acceptance, to provide an

independent assessment of the product and/or the reliability of test results obtained from process control and acceptance testing. (The results of independent assurance tests are not to be used as a basis of product acceptance.)

- *Acceptance Plan.* An agreed-upon method of taking samples and making measurements or observations on these samples for the purpose of evaluating the acceptability of a "lot" of material or construction.

To the above definitions, we will add two that are not contained in *Transportation Research Circular E-C074.* These definitions are for "quality management" and "project quality assurance." Another Transportation Research Circular that focused on DB, *E-C090,* recognized the need for new definitions for quality in DB projects, stating:

> As it relates to QA, the owner is responsible for oversight management and *a new definition of QA. This new definition includes oversight* to provide confidence that the design-builder is performing in accordance with the QC plan, design monitoring and verification through auditing, spot-checking, and participation in the review of the design. (TRB 2006; emphasis added)

Quality management is defined as follows:

- *Quality Management* (*QM*). The totality of the system used to manage the ultimate quality of the design as well as the construction encompassing the quality functions described above as QA, QC, IA, and verification.

In a report by the National Cooperative Highway Research Program (NCHRP 1979), defining a QM system was simplified to four basic questions that provide a concise reference to ensure that a QM system is functioning properly. The questions are: What do we want? How do we order it? Did we get what we ordered? What do we do if we don't get what we ordered?

Project quality assurance is defined as follows:

- *Project Quality Assurance* (*PQA*). All those actions necessary for the owner to ensure that design-builder-performed QA activities give a true representation of the quality of the completed project. This may include owner verification and acceptance testing, or IA, as owner oversight actions when the design-builder is assigned the responsibility for design and/or construction QA activities. Additionally, these also include owner oversight, verification, validation, acceptance, and other activities necessary to satisfy statutory requirements for public projects, and the employment of independent quality consultants that may be necessary in DB projects with postconstruction operations and/or maintenance options.

Design-Build Project Quality Assurance Model

Figure 4-1 is a graphical representation of the DB PQA model. The shaded area represents the universe of QA requirements that exist during both the design and construction phases of a DB project. In this form, it makes no specific assignment of QA/QC roles and responsibilities between the owner and the design-builder. The owner is free to make those assignments to whichever entity is best suited to carry them out in a satisfactory manner. The model shows that no matter who actually performs the classic design and construction QA/QC tasks, at some point the owner must make a business decision as to whether to accept the completed design product and the finished construction product. In the case of design, this decision is indicated when the owner agrees to allow the completed construction documents to be released for construction. In construction, this decision is indicated when the owner agrees to make final payment. When both of these decision have been made, the DB project is accepted. The details associated with the making of these two decisions contribute to the DB project's acceptance plan, as previously defined. For purposes of this model, the acceptance plan will use the definition proffered by Burati et al. in an FHWA report on QA specifications. The acceptance plan "will be considered to represent only those functions associated with acceptance" (Burati et al. 2003, p. 4).

Along the way, the owner may use some form of IA to provide information that will assist it in making the design and construction acceptance decisions. In design, IA could take the form of sending portions of the design to another design

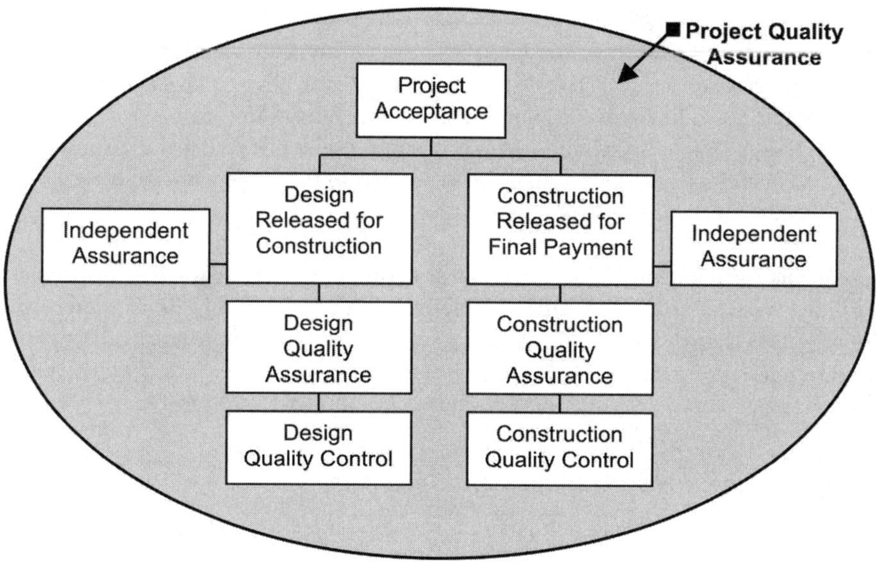

Figure 4-1 DB project quality assurance (PQA) model.
Source: Gransberg et al. (2008), Figure 3, with permission from the Transportation Research Board of the National Academies.

professional for peer review prior to releasing it for construction. In construction, IA could involve sampling and testing to statistically validate the design-builder's QA and QC testing programs. Finally, PQA also encompasses the less formal activities that the owner engages in to facilitate the design-builder's progress. DB project delivery demands a rich flow of technical information between the owner and its design-builder (Beard et al. 2001); as a result, owners have developed mechanisms to satisfy this requirement. One such activity has become known as "over-the-shoulder" design reviews (these are discussed in detail in Chapter 3). The Minnesota DOT (MnDOT) defines these as follows:

> The over-the-shoulder reviews are not hold points that restrict the progress of design. They are simply reviews of the design as it progresses and opportunities for [the owner] to provide comments and feedback on the design. (MnDOT 2005)

Another example of the types of activities that fall into the PQA universe is found in the Virginia DOT (VDOT)'s DB guide (VDOT 2007), which defines two new QM roles beyond QA and QC and calls them "Owner Independent Assessment" and "Owner Independent Validation." They are defined as follows:

- *Owner Independent Assessment.* Oversight performed by the Department (or agent) to satisfy VDOT and FHWA's requirements for documenting that proper QC and QA are being performed. This oversight provides an independent assessment of Design-Builder's implementation of and compliance with the approved Quality Control and Quality Assurance plan.
- *Owner Independent Validation.* Oversight performed by the Department (or agent). The focus of owner independent validation is to verify Design-Builder's QC and QA compliance and confirm that the quality characteristics of the products incorporated in the project are valid for acceptance and payment. (VDOT 2007)

Quality assurance in DB is currently developing. The same universe of responsibilities exists in DB as in DBB; the difference lies with who holds the responsibility. While there is no consensus among the various entities that use DB as to a recommended or preferred QA organization, all owners have shifted more of the quality responsibility to the design-builder than is typically seen in DBB.

Design-Build Quality Management Spectrum

In DB, the Request for Proposal (RFP) establishes the distribution of QM responsibilities. The owner can choose to assign specific responsibilities to the design-builder and retain the rest for itself. Additionally, it is also possible to retain a third party to conduct QM activities. This firm could be under contract to either the design-builder or the owner. The literature indicates that this third party is known by many different titles, some of the more common being "general engineering consultant," "design oversight consultant," and "independent quality consultant." It is not that the processes or activities for assuring quality are different in DB; the vast majority of activities will remain the same. The difference lies with who will be responsible for performing the activities (Tam et al. 2003). In 74% of the projects reviewed by a recent study of DB QA, QM responsibilities were at least generally assigned in the solicitation documents (Gransberg et al. 2008). This is almost identical to the study's accompanying survey response where 72% of respondents answered affirmatively to the question: Does either your RFQ or your RFP define QM roles and responsibilities? Once the QM distribution decision is made, both the design-builder and the owner must assemble organizations to carry out those functions. Table 4-1 shows typical design and construction QM tasks that must be assigned to one party or another to ensure quality on any transportation construction project.

Figure 4-2 is the theoretical spectrum of QM extending from one end, where the owner conducts all QM functions, to the other where the owner assigns the total QM program to the design-builder and satisfies its FHWA-mandated oversight responsibilities (Stefani 2004) using some form of PQA. A 2006 study completed by Potter and McMahon (2006) for the Maryland State Highway Administration

Table 4-1 Typical Quality Management Tasks on Transportation Projects[a]

	Design	Construction
Quality Control (QC)	Technical review of design deliverables Checking of calculations Checking of quantities Review of specifications	Technical review of shop drawings Technical review of material submittals Checking of pay quantities Routine construction inspection QC testing Establishment of horizontal and vertical controls on site
Quality Assurance (QA)	Acceptance of design deliverables Approval of final construction documents Approval of design progress payments Approval of post-award QC plan	QA inspection QA testing Verification/acceptance testing Approval of construction progress payments Approval of post-award QC plan
Project Quality Assurance (PQA)	Approval of post-award QM/QA plans Audit of design QA activities Over-the-shoulder design review	Approval of post-award QM/QA plans Independent verification/acceptance testing Oversight

[a] These lists are not meant to be all-inclusive.

Source: Adapted from TRB (2005).

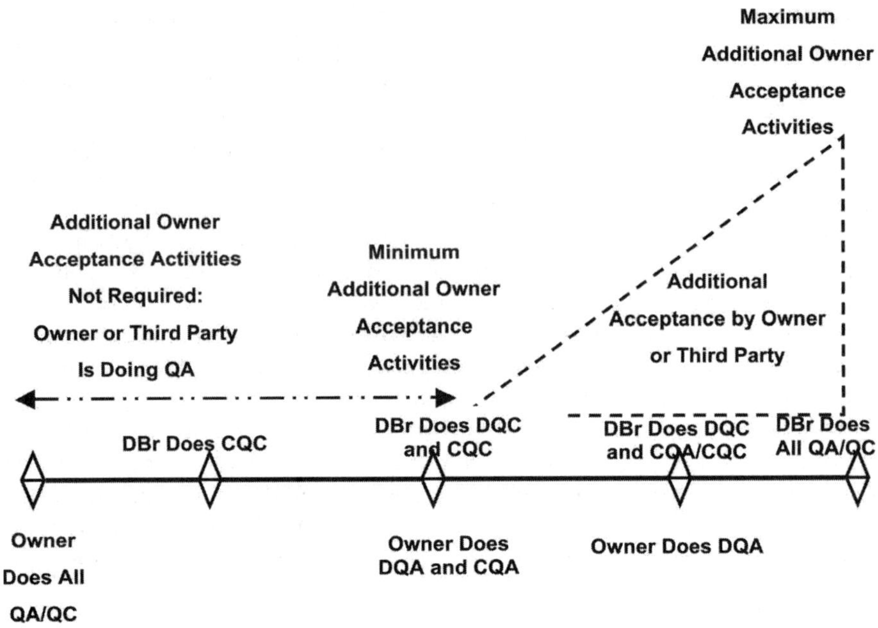

Figure 4-2 The quality management spectrum.
DBr, design-builder; DQC, design quality control; CQC, construction quality control; DQA, design quality assurance; CQA, construction quality assurance.
Source: Gransberg et al. (2008), Figure 8, with permission of from the Transportation Research Board of the National Academies.

on the subject QA and QC organizations for DB mega-projects substantiates the spectrum shown in Fig. 4-2.

Table 4-2 expands on the Fig. 4-2 concept by showing the details of the possible combinations of QM functions in tabular form. It moves from assigning all the QM functions to the design-builder, as happens in many public–private partnership projects overseas (Tyborowski et al. 1997), to the other end of the spectrum where the owner holds all the QM functions. It also includes third-party QA participation as found in the DB RFP content analysis. This results in 14 different types of QA organizations.

Applying the Design-Build Project Quality Assurance Model to the Organizations

By definition, PQA cannot be performed by the design-builder. It must be performed by the owner or by an independent third-party firm. This provides a check to the design-builder being responsible for all the QA and QC. Table 4-3 is a revision of Table 4-2 with the added PQA category in QM organizations where PQA is necessary. It shows the four most common DB QM organizations found in a recent study of DB transportation projects (Gransberg et al. 2008).

Table 4-2 Possible Quality Management Organizations

	Design QA	Design QC	Construction QA	Construction QC	Comments
Type 1	DBr	DBr	DBr	DBr	Owner oversight of design and construction
Type 2	DBr	DBr	Owner or 3rd	DBr	Owner and 3rd party share construction QA
Type 3	DBr	DBr	DBr and Owner	DBr	Owner and DBr share construction QA
Type 4	3rd	DBr	3rd	DBr	3rd party QA; DBr QC
Type 5	3rd, DBr, and Owner	DBr	3rd, DBr, and Owner	DBr	QA is shared; DBr QC
Type 6	Owner	DBr	DBr	DBr or 3rd	Owner design QA only
Type 7	DBr	DBr	Owner	DBr or 3rd	Owner oversight of design
Type 8	Owner	DBr	Owner	DBr	Owner QA; DBr QC
Type 9	Owner	DBr	Owner or 3rd	DBr or 3rd	3rd party is involved in construction QA or QC
Type 10	DBr and Owner	DBr	Owner	DBr	Owner and DBr share design QA only
Type 11	DBr and Owner	DBr	DBr and	DBr Owner	Owner and DBr share QA
Type 12	Owner	Owner	DBr	DBr	Owner oversight of construction
Type 13	Owner	Owner	Owner	DBr	DBr Construction QC only; traditional DBB QM
Type 14	Owner	Owner	Owner	Owner	In-house project

DBr, design-builder; Owner, the project's owner; 3rd, third-party (independent firm retained to conduct QA and QC or independent assurance activities); DBB, design-bid-build.

Source: Adapted from Gransberg et al. (2008), Table 3.

A 2006 Maryland State Highway Administration (MDSHA) study of DB mega-project QA and QC found that 8 out the 11 case study projects surveyed in that research used what the Maryland report called "quality oversight," which is essentially the same as the activities that this model calls PQA (Potter and McMahon 2006) and validates the above findings. It went on to say that the MDSHA "Do[es] not have a separate Quality Oversight (QO) program developed . . . so [will] need to develop the program and train their staff" (Potter and McMahon 2006).

When the owner or a third-party firm is involved in QA activities, there is no need for construction PQA activities. In Table 4-3, this is shown in Types 2, 3, and 7. PQA is needed only for the design portion of the work because the owner is involved in construction QA. Some examples are given below to illustrate how this has been defined in RFPs. Keep in mind that all these RFPs first assigned the QA responsibility to the design-builder. For Type 1, this includes both design and

Table 4-3 Quality Management Organizations with Project Quality Assurance

	Design QA	Design QC	Design PQA	Construction QA	Construction QC	Construction PQA
Type 1	DBr	DBr	Owner or 3rd	DBr	DBr	Owner or 3rd
Type 2	DBr	DBr	Owner or 3rd	Owner or 3rd	DBr	None
Type 3	DBr	DBr	Owner or 3rd	DBr and Owner	DBr	None
Type 7	DBr	DBr	Owner or 3rd	Owner	DBr or 3rd	None

DBr, design-builder; Owner, the project's owner; 3rd, third-party (independent firm retained to conduct QA and QC or independent assurance activities); PQA, project quality assurance.

Source: Adapted from Gransberg et al. (2008), Table 4.

construction QA, and for Types 2, 3, and 7 only design QA was assigned to the design-builder.

Design Quality Assurance

The design phase of a DB project is the phase where the ultimate quality of the constructed facility is quantified through the production of construction documents. A previous study of DB QM stated: "Quality cannot be assumed into the project. It must be designed and built into the project in accordance with the DB contract itself" (Gransberg and Molenaar 2004). It is intuitively obvious that the final quality of the construction is directly related to the quality of the project's design. Thus, design QA and design QC activities are necessary to assure the final quality of the products produced during the design phase of the DB project.

An important factor in achieving a high-quality design is free and open communication between all parties during the design phase (Beard et al. 2001), and the Design-Build Institute of America (DBIA)'s *Manual of Policy Statements* states: "DBIA advocates both formal and informal project partnering and considers the partnering philosophy to be at the foundation of design-build delivery" (DBIA 1998). Chapter 1 has already discussed the importance of partnering in post-award DB project administration. Success in achieving a high-quality DB project is directly proportional to the quality of the communication between the owner and the design-builder as well as the internal communications within the DB team itself. This is particularly true during the design phase.

Design Quality Defined

Transportation Research Circular E-C074: Glossary of Highway Quality Assurance Terms defines "quality" as follows:

> (1) The degree of excellence of a product or service; (2) the degree to which a product or service satisfies the needs of a specific customer; or (3) the

degree to which a product or service conforms with a given requirement (TRB 2005).

Given these three possible definitions as the basis for defining design quality, it would seem that the last one, conformance to a given requirement, best fits design quality in the DB context. The "requirement" to which the design must conform would be detailed in the DB project's RFP as promulgated in the signed DB contract. Figure 3-2 in Chapter 3 shows the hierarchy of performance elements that could be contained in a DB project RFP. One can see that it ranges from a very general "essential function" where the owner is allowing the design-builder the widest possible latitude in selecting a technical solution for a functional requirement, to a "prescriptive specification" where the owner has made the design decision and expects the DB project's design to include it precisely as specified. From the perspective of the DB design quality manager, it does not matter where on the hierarchy a given requirement may fall; it can be used as the benchmark against which the final design is checked to ensure conformance. Therefore, design quality is essentially defined on a project-by-project basis in the owner's DB solicitation and contract documents.

Owner's Role in Defining Design Quality Requirements

Owners give up control of the details of design by selecting DB project delivery. Thus, depending merely on the qualifications process to guarantee design quality, as in a pure design contract, may not be sufficient. With the predominant organizational type being a constructor-led DB team (Songer and Molenaar 1996), the designer's client is no longer the owner. Therefore, it is imperative that the DB teams' approach to producing a quality design be evaluated prior to contract award. Assuming that this has been done during the pre-award phase of the DB project, the owner must convey its contractual definition of design quality during the design phase. Often the DB RFP definition is global in nature and does not directly address specific elements of design quality that are encountered during detailed design. Therefore, the owner must continually provide clarifications to the contractual definition that are faithful to how the definition was interpreted and bid by the design-builder prior to award. For this reason, having a clear definition of design quality is imperative. If quality was not clearly defined prior to contract award, both the design-builder and the owner would have to hammer out a clear definition that is agreeable to both parties prior to starting detailed design.

The owner's role in defining design quality requirements consists of the following three major areas:

- Establish and communicate design quality objectives.
- Ensure compliance with both the contract and the implicit statutory requirements applicable to the project.
- Define the roles of the various contributors to the design.

These three areas furnish a framework within which the owner can add value to the design QM process. The major mistake the owner can make is to assume that the design-builder will "know" what the owner expects in these three areas. The best way to ensure that the quality definition is both clear and enforceable is to make it a contract requirement by publishing this information in the project's RFP and incorporating it in the DB contract. However, if this opportunity has been missed, it can be reestablished after award by ensuring that these topics are all covered in both the pre-work and the pre-design conferences, as well as addressed during the project partnering process. One should also note that if the owner does not take the initiative to fulfill this role, the design-builder needs to intervene early in the project and insist that the project's definition of quality be developed and documented, thus avoiding a potential costly dispute later in the project's design development process.

Establishing Design Quality Objectives

There are typically four categories in which design quality objectives must be established:

- Cost
- Performance
- Function
- Aesthetics

These categories represent the totality of the design-builder's design solution for the project's design problem that was articulated in the DB RFP. After contract award, the design-builder must design to both cost and schedule (Gransberg and Molenaar 2004). Thus, the details of the final design represent the only variable in the Cost-Schedule-Quality triangle. Therefore, it is important that the quality depicted in the final design be well understood by both the owner and the design-builder.

Cost Quality Objectives

The cost is fixed contractually upon DB contract award, so there would seem to be little opportunity for the owner to establish cost objectives with respect to quality. That may in fact be the case in some contracts. However, if the owner has put very little design development in the RFP, it may have relied on cost performance criteria in the RFP to reveal the design-builder's thought process in developing the proposal and to obtain a competitive breakdown of project costs to use later in change order negotiations. Hence, because performance criteria for cost have been used to form the DB contract, the owner must establish quality objectives that pertain to the RFP cost criteria, to ensure that the final design conforms to the contract. Generally, there are three types of cost criteria (Gransberg et al. 2006):

- Cost limitation criteria
- Cost breakdown criteria
- Life-cycle cost criteria

Cost Limitation Quality Objectives Quality objectives for cost limitation criteria are generally straightforward. They define the either cost constraints of the project or cost-related goals for the project. The following is a list of typical cost limitation criteria:

- Maximum price
- Target price
- Funds available
- Public project statutory limits
- Type of funding
- Multiple fund sources
- Fiscal year funding
- Owner-furnished property
- Allowances
- Real estate
- Private project minimum required internal rate of return

The cost quality objectives for maximum price, funds available, public project statutory limits, type of funding, and multiple fund sources merely consist of ensuring that the design-builder and its design team understand the limitations imposed by these criteria and then back-check to ensure that these criteria are met.

Setting quality objectives for target price criteria often gets more complicated. A target price criterion operates in much the same manner as the maximum price criterion but is less restrictive. It conveys the desired level of overall quality the owner desires, using financial rather than technical terms. Target price criteria are often stated as unit prices rather than lump-sum amounts. The owner uses these criteria to constrain the proposed design solutions to proper cost levels; to help guide the design-builder's proposal development; and to ensure that the proposed solution will be one that fits the owner's intent. These criteria all serve to make these cost limitations a part of the final contract. Here is an example of a target price criterion:

> The interior design packages for the various types of spaces in the building shall comply with the following target unit prices:
>
> Executive office spaces: $XX.XX per square foot
> Administrative office spaces: $YY.YY per square foot
> Storage spaces: $ZZ.ZZ per square foot
> All others: Not to exceed $WW.WW per square foot

Thus, the owner in this example has conveyed its expectations for the quality of interior finishes in different types of spaces in the building in a manner that will allow both the interior designer and the cost engineer to regulate the final technical design details of these features of work. Having established these criteria, the owner will now use this information to check the design packages as they are completed to see if they are within a reasonable range of the target criteria. To do this, the owner will need to communicate design quality objectives with the design team prior to the detailed design on the features to which these criteria apply. This communication will contain details of owner preferences and exclusions that may not have been contained in the RFP. For instance, in the above example the owner might indicate that it wants to exclude the use of vinyl wall coverings and it would like to explore a specific grade of carpeting in certain spaces. Knowing these design quality objectives, the interior designer can then develop options that not only conform to the owner's quality objectives but also meet the target cost limitation criteria contained in the contract.

Continuing this train of thought in a transportation example, another type of a target price criterion uses a lump-sum amount as follows:

> The landscaping around bridges, interchanges, and rest areas, including sodding, trees, and plantings, shall cost $XX,XXX ± Y% per site. The proposal shall contain a narrative describing the details of the proposed landscape plan for a typical area.

Thus, the owner in this example effectively told the design-builder the cost for a specific feature of work and asked for details about the quality that is being proposed for that fixed amount of money. The owner's objective was to get the design-builder to furnish as much landscaping as possible for the target price through competition with other offerors during the procurement phase. Having accomplished that, the owner would now negotiate the final details of this feature of work during the design phase by asking the design-builder to furnish alternatives that meet the target price criterion, among which the owner can select.

Establishing clear quality objectives is also important when a public owner has fiscal year funding limitations that will potentially constrain the design-builder's schedule. In this case, the owner must ensure that the design-builder is fully aware of these limitations and how they will eventually affect project progress. For instance, it is imperative that the design-builder be required to submit a cost-loaded schedule for the owner's approval. During the approval process the owner will check to see that the fiscal year funding limitations are not exceeded in each fiscal year of a multi-year project. If they are, it is incumbent on the owner to identify those features of work that cause the criterion to be violated and point them out to the design-builder, giving it the opportunity to revise its schedule or choose to proceed knowing that it will not receive full payment for all its work in each fiscal year in which the criterion is exceeded. Failing to do this could have a detrimental impact on design and construction if the design-builder's subconsultants and subcontractors are unintentionally put in a negative cash flow position.

The use of owner-furnished property in a project is a typical cost limitation criterion. Many public and private owners operate and maintain multiple facilities that are constantly being rehabilitated, upgraded, and renovated. The owner normally wants to use as much of the existing property as possible. However, in DB, the design-builder must commit to a firm, fixed price before the detailed design is complete. Often, the owner will not know exactly how much of the existing property can be technically incorporated in the new project. In fact, the owner will sometimes use this factor as a means of determining which proposal is indeed the best value. Thus, it becomes important for the design-builder to have clear quality objectives for those owner-furnished aspects of the project. The owner does this during the design phase in its pre-design conference by exchanging technical information on the owner-furnished property to ensure that the DOR can incorporate that property in a manner that does not degrade the intended quality of the design-builder-furnished design and construction.

"Allowances" constitute another category of cost limitation criteria for which quality objectives must be set. An allowance is usually a fixed sum of money that the owner directs to be included in the proposed price for a specific purpose within the project. These are often set to ensure that there is sufficient funding to cover the costs of construction for those things that cannot be adequately designed during the design phase. A good example of this issue often occurs in medical facilities where owner intends to install the very latest in medical technology immediately before opening a facility, which may take two or three years to design and build. At the time when the detailed design is completed, the owner cannot yet tell the design-builder the exact details of the various pieces of medical equipment that will ultimately be installed in the facility. Thus, the quality of these installations will need to be addressed during the design phase. The DOR will need to proceed with the design, recognizing that changes will need to be made after the design is complete. It can account for these by furnishing spare conduits, additional clearances, and other features that will make the final design easy to modify once the final details of the installations are known. This will keep the cost of these changes within the allowance established for their benefit.

Commercial developers who use DB to deliver their projects will often set certain constraints on the cost of the real estate that is procured in conjunction with the design and construction of the project. The purpose of these cost limitation criteria is to ensure that the land cost does not cause the internal rate of return on the development's investment to be reduced below assumptions that were used in the project pro forma. Thus, it will be important for these private owners to establish cost quality objectives for this portion of the project. For example, a description of the minimum area, maximum distances to utility service, and other technical issues that could affect the suitability of the land acquired as a part of the DB project would be communicated to the design-builder to ensure that not only the technical quality of the of the project is achieved but also that the financial objectives are met. This issue also relates to cost limitation criteria for internal rate of return that were set to ensure that the project is financially successful. This is usually done by merely establishing the maximum allowable cost of design and construction. However, this approach does not create a bias toward minimizing the initial capital costs, which would make the venture's rate of return even greater.

The owner may therefore ask the design-builder to commit to a "guaranteed maximum price" (GMP) and offer to split the savings that would be accrued by bringing in the DB project below the GMP. Thus, with a GMP and the potential for split savings added to the DB project, the owner will need to be especially careful to ensure that the technical quality is not compromised.

Cost Breakdown Quality Objectives Cost breakdown criteria are typically used to establish a means for the owner to better understand the basis of the design-builder's price proposal. As such, it would seem that their utility expires with the award of the DB contract. However, they also serve to establish the foundation on which the cost of change orders and contract modifications will be negotiated. Therefore, they can be used to set quality objectives in this realm as well. Particularly important are those that were used to constrain the financial assumptions used by the design-builder to arrive at a final price. These normally take the form of set financial assumptions used for multi-year projects where inflation as well as labor and/or material escalation might have a significant impact on the lump-sum price. Thus, the owner must remain cognizant of the effect of these directed assumptions on actual project cash flow and be prepared to adjust project cash flow if these criteria prove to be seriously out of line with the market. This is often done by establishing a management reserve fund from which short-term spikes in the market can be financed to avoid putting the design-builder in a negative cash flow position, which would inevitably threaten both technical quality and timely completion.

Life-Cycle Cost Quality Objectives One of the advantages of DB is that it allows an owner to take a longer look at the project's ultimate costs and determine life cycle cost rather than initial capital cost. Research has shown that the calculation of project life-cycle cost is a relatively straightforward application of engineering economics (FHWA 1998). Often these criteria are stated in terms of operations and maintenance requirements and attempt to influence the design-builder into selecting technical features that minimize these types of postconstruction costs. As a result, it is imperative that the owner establish quality objectives at the outset of design to ensure that what the design-builder offered in the winning proposal is indeed realized in the final design.

These objectives can be framed in terms of easily quantifiable variables, such as a maximum annual energy usage or a minimum useful life before replacement or rehabilitation. It is vital that the DOR and its design subconsultants start the design process with a clear understanding of these objectives so that the design effort is both focused and not wasted on evaluating nonviable alternatives. Again, this is a communications exercise that can be accomplished during the pre-work and pre-design conferences and amplified during design charettes and design reviews conducted during the design phase of the project.

These objectives are particularly critical if the project is being delivered with postconstruction operational and/or maintenance features. This typically occurs in design-build-operate-maintain (DBOM), design-build-operate-transfer (DBOT), and in many public–private partnership (PPP) projects. In these types of DB projects, the owner must develop quality objectives for the postconstruction period. Cost quality objectives are among these. These are often quantified in the price proposal in the form of a payment schedule for operations and maintenance.

Performance Quality Objectives

In our previous book we developed the following definition for a "performance criterion":

> A rule by which the effectiveness of an operation or function is judged and its value measured. (Gransberg et al. 2006)

So, keeping in mind that the operative words in the definition are "effectiveness of operation or function," "judged," and "measure of value," the owner can now establish a set of quality objectives that correspond with the performance criteria contained in the DB contract and articulate the "rule." A 2001 study found that "proper documentation of customer requirements was an important contribution to improving the quality of design" (Pheng and Yeap 2001). This particular study dealt with applying to DB construction projects the theory of "quality function deployment" (QFD), which was developed for manufacturing processes. According to Sui Pheng Low, a prolific author in this field, QFD is a "quality improvement technique that deals with quality problems from the outset of the product design and development stage and assures that customers' requirements are accurately translated into appropriate technical requirements and actions" (Low and Yeap 2001). Thus, when the owner (which, in QFD model terminology, is the "customer") establishes performance quality objectives in a DB project, it is in fact documenting its requirements in a manner that should ultimately enhance the quality of the project's design.

It would be both redundant and unproductive to attempt to detail all the types of performance quality objectives that could be established in a typical DB project. Therefore, the reader is referred to Chapter 3 of our previous book for the details of typical DB performance criteria. There are essentially four types of performance criteria that must be developed for every DB project: management, schedule, technical, and cost. It suffices to say that there must be a singular quality objective for every performance criterion.

The first, and perhaps most important, design performance objective defines the quality of the design submittals themselves. The owner must furnish the design-builder the details of how it needs the design product to be packaged to permit the owner's design representatives to perform the design QM activities during the design phase. Research has shown that the primary reason why owners select DB project delivery is to compress the project's schedule (Songer and Molenaar 1996), and the major milestone that influences the project's timely completion is the approval of the final design (Beard et al. 2001). Thus, if for no other reason than to expedite the review and approval of design submittals, a clear definition of their content, packaging, and level of detail is in order. In fact, at the pre-design conference the owner and the DOR should negotiate the details of each and every design submittal to ensure that the project is not delayed due to lack of the necessary design documentation for the owner to make timely approval decisions. This is especially important if the contract calls for the design-builder to utilize building information modeling (BIM) or some other multidimensional design process. Using the performance criterion definition, this is an example of how the "effectiveness" of the design process will be "judged."

One technique to simplify this process is to break the project into disciplinary design packages rather than the traditional submittals based on an arbitrary percentage of design completion. Disciplinary design packages line up nicely with the design subcontracts and, as a result, they further line up with the technical disciplines the owner will need to review them. Thus, civil engineers review design packages that are primarily civil engineering, rather than trying to extract the civil engineering portion out of the greater project design package. This technique also allows the owner to release stand-alone design packages for construction on a package-by-package basis, making payment for design progress straightforward and unambiguous. In other words, instead of trying to estimate what percentage of the structural engineering is complete from within a 60% design submittal, the owner can pay 100% of the structural design cost when the structural design package is found to be complete per the defined quality objective and approved for construction.

Once the form and format of a design submittal are established, quality performance objectives can be determined for individual features of work. For instance, if one of the owner's overall project objectives is to achieve as high a level of sustainability as possible within the constraints of budget and schedule, it could instruct the DOR to search for sustainable options wherever possible, and make material and process selection decisions by picking the alternative that would be awarded the highest number of points using a commonly acceptable scale, such as the U.S. Green Building Council (USGBC)'s Leadership in Energy and Environmental Design (LEED) certification process. Thus, while there may be no contract requirement to achieve LEED certification, the owner has clearly articulated an interest in minimizing the building's environmental impact and has given the design-builder's team a metric against which to measure possible alternatives that conform to the contractual budget and schedule constraints. By doing so, all the designers and constructors on the DB team understand the performance quality objective for sustainability. Returning to the performance criterion definition, this is an example of how the "value" of a given design alternative will be "measured," and it uses the LEED standards as a "rule."

Another example of a performance quality objective would entail the owner of a highway rehabilitation project indicating specific dates on which it would prefer that traffic flow through the work zone be maximized. This type of quality objective is not contractual. Nevertheless, the owner is sharing its knowledge of local traffic conditions with the design-builder in advance of the traffic control design in a manner that is mutually beneficial to both parties. Thus, the traffic control plan can account for these dates and the construction can proceed with a plan to minimize disruption for discrete periods of time. In fact, in some major DB projects, the design-builder has used this type of input to develop work zone traffic flow computer simulations that assist it in optimizing the project's traffic control plan. Again, according to the performance criterion definition, maximizing traffic flow on specific days forms a "rule for measuring the effectiveness" of the final traffic control plan.

A third typical example would involve an owner of a wastewater treatment project describing to the design-builder the requirements for the treatment facility to be able to increase its output as the population it serves grows over time. It is extremely important for the DOR to understand, at the outset of design, the need to eventually expand the project's footprint and add additional capacity. Establishing capacity expansion potential as a quality performance objective will have a huge impact later in the project's life cycle, when the owner needs to engage another design-builder to actually undertake the expansion when the demand reaches the level that justifies it.

All three of the above examples demonstrate the need for free and open communications during the design administration process. QM is most effective when it is practiced in a business environment where both parties feel that they can be honest and discuss bad news without fear of retribution. The ultimate goal is for both the owner and the design-builder to take those actions during the design phase that furnish the desired performance within the cost and schedule constraints of the DB contract.

Functional Quality Objectives

Quality objectives for function differ from those established for performance in that they focus on *what* must be included in the design rather that *how* a given feature will work. Thus, functional requirements are often expressed by the implicit requirement to design a project in accordance with existing codes, regulations, and laws. The owner does not need to specifically cite a given building code to enforce its application to a DB project. Nevertheless, many codes give the designer some latitude in how they can be met, and within that latitude the owner may have preferences among the applicable options. Thus, expressing these preferences constitutes setting a "functional quality objective."

For instance, the plumbing codes in some areas allow mechanical engineers to use other methods than P-traps on drain lines. If an owner's experience has been good with P-traps, it can set a functional quality objective by limiting the mechanical design to the use of P-traps. This type of early communication is important to achieving the required design schedule in that it eliminates controversy over allowable technical alternatives with which an owner has no experience and, as a result, it deems to be riskier than tried-and-true design solutions. This understanding is established at a point in time where no design effort has been expended. One word of caution is in order, though. If the various alternatives to a given code requirement carry significantly different prices, the owner should either publish its preference for a given alternative in the RFP or be prepared to execute a change order if the design-builder had assumed a lower-cost alternative in its price proposal.

A second example of a functional quality objective deals with the circulation of a building's occupants. The owner in this case would furnish the DOR a list of space functions that need to be in close proximity and then allow the DOR to design a floor plan that optimizes the requirement. This essentially forms the "rule" for this aspect of building "functionality." Many architects would argue that this is merely part of the normal design process, and in most cases it is. However, it is presented here as an example of how the owner set a performance quality objective for something that is important to the ultimate product but is not specifically articulated in the technical portion of the DB contract.

A final functional quality objective example comes from the transit industry. Transit system owners have a strong financial interest in the efficient functioning of their trains, and commonly promulgate functional quality objectives to support this. For instance, the ability of the transit system's passengers to board and disembark from the train cars can drive the design of the transit system's stations and platforms. Additionally, accommodations for passengers with disabilities must also be factored into transit station design at an early point in the design process. These types of functional quality objectives are often quantified in terms of "headway," or the number of minutes between trains in the system. Thus, the DOR must understand the standard for the system's function in terms that both capture the essential physical constraints and the financial requirements without constraining the possibility for creative solutions to this given design problem. Thus, the transit platform designers know that they must include all the features necessary in the platform to allow all types of passengers to be able to embark and disembark within the time parameters established by the headway for a given train.

Thus, it can be seen from these three examples that functional quality objectives are focused on what must be designed without prescriptively detailing the design solution. It is important that these objectives be both clear and unambiguous to both the owner and the design-builder. It is also extremely important that they be established as early in the design process as possible to ensure that the designers do not waste design effort generating solutions that do not measure up to these objectives.

Aesthetic Quality Objectives

"Aesthetic quality objectives" are inherently subjective and, as a result, require that both parties carefully communicate the details for meeting these throughout the design process. Like the cost, performance, and functional quality objectives described above, aesthetic quality objectives may have been established in the project's procurement documents and proposal, but nevertheless must be addressed and well understood at the outset of the design process. If aesthetics were included in the evaluation criteria and used to make the best-value award decision, the owner must review, with the design-builder's design team, the proposed design from an aesthetic point of view and indicate those features that are important to the owner and should not be altered, as well as those that are of little value or to which the owner is indifferent. By doing so, the DOR is given the greatest possible latitude to balance the important aesthetic qualities of the project with its technical requirements, and is not unnecessarily constrained by trying to maintain features that are not important to the owner's desired view of the project's appearance.

The simplest type of aesthetic quality objective is one that asks the design-builder to match the look of existing structures in the area where the project is being constructed. This quality objective may seem relatively straightforward, but the operative word is "match." This can be interpreted to mean as little as "Make sure that the exterior finishes are similar" or as much as "Use exactly the same type and color of exterior finishes." For example, in the first case the owner is merely stating that it doesn't want a modern-looking structure in juxtaposition with the traditional structures that surround it. In the second instance, the owner has become quite prescriptive and the design-builder's design team will need to be very careful to not violate this specific quality objective. Again, it is evident that the timely communication of this simple aesthetic quality objective is paramount to its achievement.

A second example of an aesthetic quality objective concerns the relative quality level of interior finishes. In a large corporate office building, for instance, one would find a number of different rooms that serve different functional purposes. Each room will have a different requirement for aesthetic quality. The owner in this case could establish an aesthetic quality objective for different levels of finish treatments by grouping similar functions together and creating a definition of the aesthetic expectations for this "group." For instance, executive office spaces would be assigned one quality level, whereas utility and storage spaces would have a different requirement. These objectives would be further defined through the submittal process, using color boards and samples as the media for communicating design intent.

Aesthetic quality objectives do not only apply to architectural projects; they are often applicable to transportation and other engineered projects as well. Most major DB bridge projects, such as the Arthur Ravenel Jr. Bridge over the Cooper River in South Carolina, are required to make an architectural statement as well as serve the functional requirement that justified the project in the first place. Additionally, many state DOTs have created aesthetic requirements for such utilitarian features as retaining walls, viaducts, and interchanges in an effort to beautify the right-of-way. Thus, an aesthetic quality objective for a DB project to rebuild an interchange in Arizona might require the design-builder to utilize a "Southwestern" theme for the colors and any patterns that may be used in the design.

In summary, establishing quality objectives for cost, performance, function, and aesthetics sets the stage for the remainder of the design QM process by articulating the owner's expectations for these areas at a time that allows the design-builder to proceed with the design development, keeping these objectives at the forefront of the DOR's process. Failing to establish these at the beginning of the design phase, or establishing them late in the process, risks misunderstandings and potential disputes. The pathway to high quality is paved by clear communications, and the owner must ensure that it starts the design phase by establishing a thorough understanding of the project's quality objectives. If the owner does not do this, then the design-builder needs to elicit this information before embarking on the design.

The Economics of Quality in the Design Phase

The economics of quality in the design phase can be best described by one of Murphy's laws: There is never enough time to do it right the first time but there's always time to do it over.

One major financial issue for the DB project's design team is minimizing wasted design effort. Design is inherently iterative, and the design team will expect a certain degree of redesign effort in every project as various disciplines are coordinated with each other and conflicts between disciplinary design products are eliminated. Nevertheless, the DB proposal never contemplates redesign because the design team failed to achieve the contractual level of quality in its design product. Therefore, it is important to ensure that both the owner's design reviewers and the design-builder's design team understand the requisite level of project quality, so that once the designers achieve that level the owner will accept it and release the design for construction.

Changing Traditional Attitudes toward Design Quality

Attaining this goal means both the owner and the design-builder must put aside several old stereotypes regarding the QM process. The first is that higher quality will lead to increased cost. DB is a project delivery process where the design-builder must commit to a firm, fixed price before the design details have been developed. Therefore, the design-builder must "design-to-cost" in order to realize its target profit for the job, and it is important for the owner to remember that once the contract is signed, the level of quality is fixed along with the price. The design-builder retains the risk that the design it proposed will fulfill all the quality requirements contained in the DB contract. However, the level of quality cannot increase above that contemplated in the contract without compensation. On the other hand, the design-builder must remember that improved design quality actually saves money and potentially increases future business with a satisfied client. Thus, it is in both parties' best interests to ensure that the required quality for both the intermediate design deliverables and the final construction documents be agreed upon before design effort is expended to advance the design.

Figure 4-3 illustrates the economics of quality in a DB project.

Another issue that must be put to rest to successfully design a DB project is the DBB notion that identifying problems with the QM process only leads to blame, faulty justification, and excuses, and therefore it is best to keep them secret. DB project delivery is inherently a cooperative effort between the owner and the design-builder (Beard et al. 2001) and requires a higher level of mutual trust and professional collaboration during project execution (Gransberg et al. 2006). Thus,

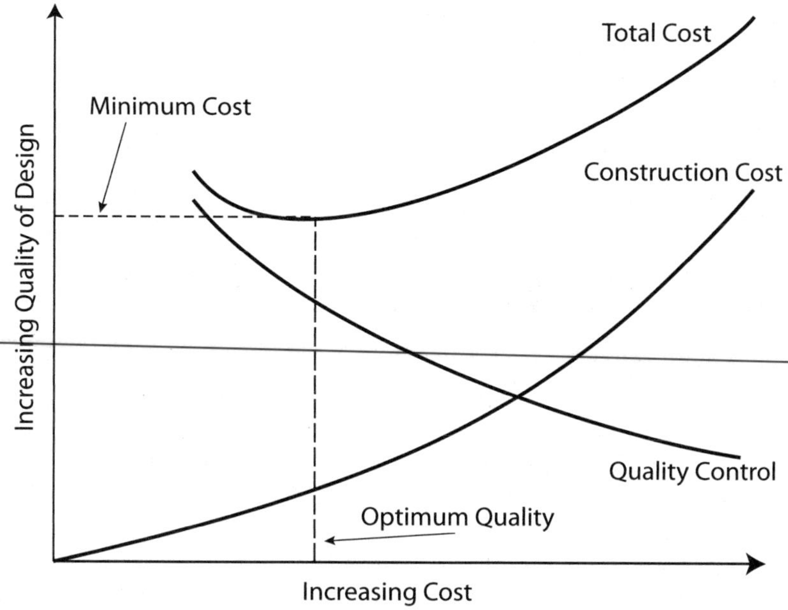

Figure 4-3 Economics of quality in a DB project.
Source: Adapted from Kirkpatrick (1970).

the owner and the design-builder must adopt an attitude that identifying quality problems should lead to cooperative solutions which not only enhance the overall quality of the project, but also of the QM system that controls both design and construction quality during project execution. Thus, to achieve the target profit for the design effort, the DB team in fact invests in a sophisticated QM process that identifies problems at the earliest possible point so that the redesign effort is minimized. Additionally, by partnering effectively with the owner's design reviewers, these problems are disclosed and jointly analyzed to determine their root causes so that any systemic defects can be immediately eliminated, regardless of their source. Again, this type of "open books" approach to DB design quality fosters the mutual trust necessary to facilitate resolution to the inevitable differences of professional opinion as to the adequacy of the design details that always occur during the review of design deliverables.

Both the designers and the builders on the DB team must always remember that quality is not internally focused. In fact, it should be customer focused. The team's ability to win the next DB project advertised by the current owner is directly related to their performance on the current one, as measured by the owner's opinion of the quality of both the design and construction effort. Thus, while the discovery of a design error or omission during design QA or design QC might ultimately result in an expensive process to rectify it, the fact that the problem was discovered demonstrates that the design QM process is effective. The design-builder's willingness to take immediate action to correct its own errors will ultimately enhance its reputation within the owner's organization, making it more likely to be successful in winning future business from this particular client, as well as making this client a good reference for future work with other owners. Thus, the DB team has an opportunity to turn a negative experience into a positive experience by utilizing the channels for free and open communications with the owner established during the initial design meeting.

The final DBB attitude that must be eliminated is that quality only occurs during project execution, specifically during construction. Ultimate quality begins at project initiation and must be planned for the project's life cycle to include the postconstruction operation and maintenance phase. The design decisions made during the opening stages of the DB project ultimately determine how well the project performs its intended purpose. Therefore, an effective design QM process is arguably more important that the construction QA/QC process. A recent study of 75 public DB projects in 35 states worth a total of more than $700 million found that only 18% of the projects required the design-builder to submit a design QM plan prior to award of DB contract, whereas 100% asked for some form of a construction QM plan (Gransberg and Windel 2008). The study concluded that owners were not recognizing the importance of design QM. This may have been due to the fact that many public owners produced their own design in DBB projects and, as a result, their design QM policies were less mature than similar policies in construction (E. DeCorso, 2004, unpublished honors thesis, University of Oklahoma, Norman, Okla.: "How Owners Communicate Quality in Public Sector Design-Build Requests for Proposals").

Design-Build Design Quality Issues

Obviously, a section on DB design quality issues could be infinitely long if we tried to cover each and every aspect of all types of projects. That is not the intent of this section. It will approach this topic on a global basis, addressing the major general issues that we have encountered in both the literature and our personal DB project experience. These are:

- Unintentional owner assumption of design liability
- Allocating QA responsibilities during design
- Resolving professional differences of opinion during submittal reviews
- Level of design detail in required design submittals
- Scope creep
- Actions that can be taken by the design-builder after submittal reviews
- Ability to change previously approved design submittals

Unintentional Owner Assumption of Design Liability

Because DB places the burden of developing the design details on the design-builder, the owner must be quite careful to define all the constraints that may exist for a given feature of work without becoming overly prescriptive. The ownership of design liability will be allocated through the owner's RFP performance criteria. The unwritten rule of thumb in this case is:

> Whoever designed it is responsible for its ultimate performance.

Therefore, the owner needs to carefully write the project performance requirements and the performance criteria that are responsive to those requirements.

Referring again to Fig. 3-2 in Chapter 3, which shows the hierarchy of performance/design elements, many owners and members of the DB industry misuse the term "performance specification." The word "specification" implies that a design decision has been made based on the professional design process. When an owner formulates a design requirement that has more than one technically feasible solution, the proper term would be "performance criterion" rather than "performance specification" because the owner has conducted a project scoping process and determined the requirements that the design-builder will have to achieve to deliver a responsive project. The requirements of that project are then articulated in performance terms, but the owner has not completed the design process and therefore is not writing a "specification." It is important to be clear in the distinctions between the two. If a project performance issue arises during or after construction, the liability will rest with whichever party actually "designed" the feature of work in question. By calling the RFP "performance criteria" "performance specifications," the owner risks confusing the legal situation with inaccurate use of terminology.

Thus, the task during the design phase is for the owner's design reviewers to be able to differentiate between those design features for which the owner has specified a given design solution, and those for which the design-builder has been given contractual responsibility for making the ultimate decision on the details of design. When a designer reviewer decides to reject a design-builder's solution, that reviewer must be able to prove that it does not conform to the applicable performance criteria. If it does conform, the rejection could be grounds for a compensable change to the contract. Taking it to the next level, the reviewer who not only disapproves the detail but then directs how the detail must be corrected risks assuming liability for performance of the directed change. As shown in Fig. 4-1, the owner's major risks lie mainly in the project approval process. In DB, the owner must also bear the risk that its personnel are going to be technically competent enough to make informed technical decisions as they arise. This is not to say that an owner's approval or acceptance of the design-builder's final design in any way relieves the design-builder from the total design liability. The issue is more subtle and goes back to the "Whoever designed it is liable for it" principle discussed above. When the owner becomes directive in its design review comments, it flirts with assuming the liability for those design directives.

For example, an owner's engineer who reviews a specification for guard rail rustproofing and doe not like the DOR's method might make the following comment:

> Change rustproofing specification to require galvanizing in accordance with ASTM standards.

By articulating the concern in this manner, the owner is usurping the DOR's responsibility and making the design decision on this feature of work. Thus, if the design-builder complies and the coating later fails to work as desired, the owner will be found to have transferred the risk for rustproofing performance to itself. However, the owner's engineer could express the same concern by saying:

> The specified rustproofing method does not have a good record of performance in this application. Reconsider the use of this method and explore other options such as galvanizing per ASTM standards.

The owner has now left the design decision squarely on the design-builder's back and furnished a clear indication of owner preference in this area. If the design-builder decides to use galvanizing and it subsequently fails, the design liability will remain with the design-builder. This is because the design review comment was *advisory* rather than *directive* in nature.

In DB, "tacit approval" is even more abstract in its application. The issue of tacit approval often arises when the owner does not know how to respond to a constructor-initiated request for approval and, as a result, does not respond at all. There is plenty of case law regarding this subject and all of it indicates that the "do nothing" option will always be construed against the owner. In DB, the possibility for tacit approval becomes even more confused, especially during the design phase of the project. If the DOR is maintaining the necessary level of free-flowing communications, there will be ample give-and-take regarding potential acceptability of various design details, especially if the project must achieve a very aggressive delivery period. There is no single method to protect an owner against tacit approval problems during the design phase except to say that owner's personnel must always respond to design-builders' requests for approval in a timely manner. That response could be as simple as to state that the matter is being taken under advisement and that this response should not be construed as approval. This ultimately requires the owner to respond in a timely manner.

In the past decade, a controversy has developed regarding the owner's role in approving the design-builder's design product. Some argue that to formally approve the final design is to return to the risk distribution inherent in the DBB process, and that "approving" the design is tantamount to warranting that it complies with the contract and thereby relieves the design-builder of responsibility to correct design errors and omissions at its own expense. Those who adhere to this school of thought prefer to use the term "accept" in place of "approve" to indicate that the owner is satisfied with the design product but will not take responsibility for its quality. On the other side, the counter-argument avers that when the owner reviews the design product and gives the design-builder authority to proceed with construction, this process constitutes an approval of that product and carries no more import with regard to transfer of design liability than approving a construction contractors' shop drawings in a DBB contract.

We have no opinion with regard to which school of thought on owner design approval is correct. The model shown in Fig. 4-1 is based on the idea that DB PQA has two major decision points, and the "release for construction" decision indicates acceptance of the final design. Therefore, this model, which was originally developed for the transportation industry, tends to support the second school of thought. However, the discussion is offered here to alert both owners and design-builders to the issue and to stimulate discussion within each organization to ensure that worries about inadvertent transfer of design liability does not lead to inaction and potential transfer of liability via tacit approval.

Quality Assurance Responsibilities during Design

One of the advantages of DB is the opportunity for the owner to contract out QA and QC activities and thus reduce the workload on the owner's employees. In a previous study of DB transportation project solicitation documents, nearly all the analyzed RFPs allocated the design QA functions to the design-builder, with the owner merely retaining oversight in the form of review and verification of the design's ability to meet the stated contract requirements (Gransberg et al. 2008). In New Mexico, the QA roles of the owner (in this case, the New Mexico DOT, NMDOT) and design-builder were defined in one RFP as:

- Owner's QA Responsibilities: Oversight and audit of Contractor design and construction, including verification sampling and testing and independent assurance.
- Design-builder's QA Responsibilities: The Contractor will be required to plan, implement and provide a Quality Assurance/Quality Control (QA/QC) Program for its design and construction operations. (NMDOT 2001)

NMDOT did, however, include this paragraph in the RFP to allow additional flexibility if it deemed it to be required:

> The Department may establish and maintain its own quality assurance and/or an independent quality assurance organization to oversee and/or perform quality audits of the Contractor's management, design, construction and maintenance activities, the Contractor's Quality Assurance procedures, Verification Sampling and Testing and the quality of the final product. (NMDOT 2001)

This caveat provides the owner with a contractual safeguard to use if the design-builder does not follow approved QM procedures. Thus, the assignment of design QM responsibilities becomes an exercise in risk analysis and management, with the owner seeking to optimize the ultimate allocation. A finding in an earlier study indicated, "Agencies that have experienced quality problems on [prior] projects are retaining QA responsibility [for themselves]" (Gharaibeh et al. 2005, p. 96). This experiential adaptation of the QM system springs from the legal issues associated with the design process, where owners are being careful to not unintentionally assume design liability by involving themselves too deeply in the design process.

Use of Internal Personnel versus Consultants Another issue that surfaces with design QA is deciding whether to perform the design QA with the owner's own design professionals, or whether an independent consultant will be retained to perform it on behalf of the owner. This issue is usually most applicable to the public sector, but many large corporations have robust facility engineering organizations to which the following discussion will apply as well. For example, the FHWA Eastern Federal Lands Highway Division (EFLHD) contracts out all of their QA while retaining the responsibility to ensure that the construction conforms to the requirements. This excerpt from an EFLHD RFP illustrates this approach:

> The Contractor [the design-builder] shall be responsible for all work as described in these RFP documents. The scope of work includes . . . quality control/quality assurance for design and construction, materials sampling and testing. . . . EFLHD will perform management, design, and construction oversight activities of the Contractor's operations and end products to satisfy the Government that the Contractor meets the contract requirements. Included in the oversight activities will be design reviews, construction acceptance, independent verification testing activities, and oversight of maintenance of traffic and permit compliance as outlined in this RFP. (EFLHD 2001)

Department of Transportation Design-Build Guideline Approaches for Design Quality Assurance Our previously noted study of DB QM practices in transportation reviewed the DB project administration guidelines published by various state DOTs. We found that there are three different policies established in determining the design QA roles in DB projects:

- Variable assignment of design QA responsibilities on a project-by-project basis.
- Assigning design QC to the design-builder and the design QA to the DOT.
- Assigning design QA and QC to the design-builder with the DOT performing oversight and verification (i.e., design PQA) (Gransberg et al. 2008)

In the first approach, the assignment of design QA roles can be varied from project to project. This recognizes that every project is different and that, depending on size, delivery speed, and technical complexity, the optimum assignment of QA responsibilities will be different based on individual project needs. This is shown in the 2006 Arkansas State Highway and Transportation Department (AHTD) *Design-Build Guidelines and Procedures*:

> The D/B package shall address any quality assurance requirements that the selected firm must follow in addition to those already in the referenced specifications, policies and procedures that will assure quality products (plans, materials, construction, etc.). QM criteria require at least three independent roles, including (1) quality control by the selected firm, (2) acceptance or verification by the Department's Resident Engineer (RE) office, and (3) independent assurance by the Department's central office staff. The responsibilities for all three roles and minimum sampling, testing and inspection frequencies shall be defined in the scope. If any of the three roles is eliminated, project quality shall be closely monitored and an objective analysis shall be made of the impact of the change on the quality of the project. (AHTD 2006)

In the second type of design QA approach, the design-builder is responsible for the design QC and the owner is responsible for the design QA. This approach parallels the DBB assignment of responsibilities for construction QM. The Colorado DOT, the Massachusetts Highway Department, and the Florida DOT use this method, and an example of this comes from the 2001 Arizona DOT *Design-Build Procurement and Administration Guide*:

> The Design-Builder shall be required to submit a design quality management plan which describes how the Design-Builder will control the accuracy and completeness of the plans, specifications, and other related design documents produced by the Design-Builder. . . . ADOT will still retain a quality verification role as it does for other quality management issues. For design work, quality verification will be accomplished through the use of design reviews led by the PM and performed by ADOT's technical groups or the general consultant, if one is used. (ADOT 2001)

Finally, the third approach assigns the design-builder both design QA and QC. The owner steps back from active participation and responsibility and, instead, only performs oversight and verification of design quality. This approach can best be called "design PQA" per the definition given in the previous section. This is followed by the New York State DOT (NYSDOT):

> The contractual requirements for design management and QA/QC are the primary responsibility of the Design-Builder rather than the Department. . . . The Department's project staff Oversight role during design and Design Review consists of monitoring and auditing design progress, interpreting contract requirements, and verifying design compliance with contract requirements. (NYSDOT 2005)

Regardless of how the design QA and QC responsibilities are assigned, they must be performed. When the owner will perform the design QA activities, the contract documents (i.e., the RFP) need not further explain design QA activities. However, when the design-builder is assigned the design QA responsibilities, it is imperative to lay out the requirements in the RFP so that confusion is eliminated or minimized and the owner understands exactly what design QA services it will receive with the proposal, as well as how to integrate its PQA activities during design.

Resolving Professional Differences of Opinion during Design Reviews

Probably the most thorny and often-encountered issue regarding design QM is the good faith difference in professional judgment that occurs when one competent design professional reviews the work of another. In our previous book (Gransberg et al. 2006), we advocated that this issue be first addressed during the procurement phase, with the owner essentially mandating and accepting some liability for those details that it believes have only one technically acceptable design solution. Thus, a comment regarding good design and engineering practice must be made at this point. If the owner has historically used a certain set of design details, such as design factors of safety in its past DBB projects, and believes that using those factors of safety are essential to a successful design on a DB project, it should not assume that any other practicing engineer will automatically choose those factors of safety. The RFP should contain a listing of the minimum required design factors of safety for every element of design that the owner cares about. The DOR in a DB contract will, by definition, be a qualified design professional. If that were not the case, the owner could not award the DB contract to that DB team. Therefore, because the design-builder will be assuming responsibility for both design and construction, the owner will have to accept that any factor of safety chosen by the design-builder's design team will reflect good engineering practice as evidenced by the designer's seal on the plans and/or specifications. If the owner elaborates its approved design factors of safety in the RFP, potential conflicts during the design phase when one competent professional disagrees with the professional opinion of another can be avoided.

On the other hand, it is impossible for the owner to elaborate every potential preference or detailed expectation that it may have on most complex DB projects. Therefore, after the project has been awarded, one item on the agenda of the initial design meeting should be to discuss how the owner and design-builder will resolve these issues if they arise. It is important that this discussion takes place *before* there is a dispute and team members take sides, and that it uses the DB contract as its foundation. An outline for the agenda for this important discussion might be:

1. Break the design scope into features of work with a great deal of specificity in the RFP, and also identify those features of work where the design-builder has design flexibility.
2. Cover the features with either limited or no options for variance during the design process so that the DOR and its design subconsultants

understand the contractual standard that is required for each of those features. Specifically cover any design criteria that might have been incorporated by reference in either the RFP or the final contract.

3. Cover the features whose design is contractually open to interpretation. The design-builder will have made design assumptions during the proposal preparation phase that affect its price proposal, and should disclose those assumptions to the owner's design reviewers to ensure that there are no significant issues inherent in the assumed design solution for those features of work.
4. Agree on a detailed list of preliminary design solutions for every feature of work in the project's scope.
5. Discuss the owner's review process for design submittals and establish a communication system whereby the design reviewer can gain information from specific design professionals on the DB team. In this, include provisions for the owner to furnish nonbinding preferences to the design team that can be incorporated as the budget and schedule constraints permit. Additionally, cover those design elements, such as paint color selection, on which the owner must be consulted before incorporating them in the design submittal.
6. Depending on whether or not a project partnering agreement has been developed, develop a system whereby a difference of professional judgment that is not clearly covered by contract language can be quickly solved. If the project is partnered, this may already have been developed.

The bottom line in this issue is pretty straightforward. The design-builder must comply with the terms of its contract. However, by requiring the design-builder to commit to a price and a schedule before completing the project's design, the owner must accept those elements of the design-builder-produced design that meet the design criteria regardless of the professional opinion of the design reviewer. If the reviewer believes that the project cannot be safely or properly constructed as designed and the design submittal *complies* with the contract, then the owner's design criteria are flawed and the design-builder is probably due a compensable change order to revise both the design and the construction.

Level of Design Detail in Design Submittals An owner that is new to DB project delivery is often shocked that the level of design details produced by an experienced design-builder is less than that seen in DBB construction documents. There are a number of good reasons for this, and owners new to the process should carefully analyze the types and quantity of specific design information its reviewers will require to be comfortable that the design accurately portrays a constructed project with which the owner will be ultimately satisfied. The issues at work here are:

- The design is a deliverable that flows out of the DB contract; it is not a part of the contract itself, as in DBB.

- The designer now works for and with the constructor and no longer needs to describe in great legal detail all the details of design for contract administration purposes. The two entities can literally discuss design issues and make decisions on the project site, if necessary, without involving the owner.
- Construction contractors do not need the level of detail typically expressed in a DBB set of construction documents to adequately build the project. Many, if not most, construction processes are repetitive from project to project and, as a result, constructors do in fact know how to build the project they have been selected for. In fact, in DB, the qualifications-based selection of constructors virtually ensures that the construction contractors will have successfully completed past projects of similar scope, complexity, and magnitude.
- Because design decisions are made by the DB team in response to performance and design criteria during proposal preparation, there is no need to prepare a detailed set of specifications. The design-builder can literally use a catalog cut (similar to a construction submittal in DBB) to replace the specification because it knows exactly the type, make, and model for most major features of work that affect the price on the day it submits its proposal.

There are no doubt a number of other reasons for the high level of detail in DB design that apply to specific projects. The upshot here is that the owner must remember that the appropriate level of design detail contains sufficient information for the constructor to properly build the feature of work in question. Additionally, the owner's design reviewer is no longer doing a technical review of the design deliverable, the way it might be used to doing on a part of a design contract in a DBB project. It is reviewing the design submittal for compliance with the contract criteria, not checking the engineer's calculations. Therefore, the appropriate level of detail from the reviewer's perspective is that which provides sufficient information for the reviewer to be able to make the compliance decision. This can be augmented with an informal chain of communications that allows the reviewer to communicate with the appropriate design professional during the review process to amplify those areas where insufficient information is provided.

Scope Creep

"Scope risk" denotes all those areas that define the technical scope of the project and the ability to confidently quantify the level of effort required to satisfy the required level of technical functionality in the finished project. The DB industry uses the term "scope creep" to express this type of risk. Scope creep happens during design development when the owner demands that work be incorporated into the final design that was not contemplated during the preparation of the project's price proposal. Thus, it is often the result of the design QM process. Also, scope creep is sometimes synonymous with "reduced profitability" because the design-builder will eventually use the contingencies built into the price proposal to account for the cost of unknown work, and when the contingencies are gone, the costs will eat into the project's profit margin. Therefore, a DB project with a solidly defined scope of work will be less susceptible to scope creep than one whose scope of work is vague. The design-builders proposing on the well-defined project will be able to minimize the amount of contingencies and propose a competitive profit margin. This will give the owner a lower price for the work. In the second case, the design-builders will have to increase not only the contingencies associated with the vague scope of work but also their profit margins to try to ensure that the project will be profitable despite the scope risk. Predictably, this will drive the proposed prices to potentially unreasonably higher levels.

This is where enlightened risk management during the procurement phase plays a strong role. The owner can choose to share the risk as a means of reducing the project's cost. Following the principle that risk should be assigned to the party that can best manage it, the owner should assume those risks that it can best manage. For example, a project whose subsurface geotechnical conditions will be unknown at the time of DB contract award leaves the owner with two options by which to distribute the scope risk associated with the design and construction of the foundation:

- The owner could place the risk on the design-builder by directing it to conduct a geotechnical survey and produce a foundation that is designed and constructed in accordance with the requirements of the final geotechnical report, including the cost of the foundation, in the lump-sum price proposal.
- The owner could share the foundation's scope risk with the design-builder by offering to pay for the foundation on a unit price or cost-plus basis and the remainder of the project on a lump-sum basis.

In the first case, the owner is shedding all the scope risk with regard to the ultimate cost of the foundation and forcing the design-builder to account for it in some fashion inside the price proposal. If the actual conditions mirror the worst possible geotechnical case, then the owner may actually benefit from having a foundation that probably cost a bit more than the design-builder estimated. However, if the opposite is true, the discrete cost of the foundation will be considerably higher than what the owner would have had to pay in a DBB contract.

In the second case, the owner forms the contract regarding the scope of foundation work in a manner that permits the design-builder to get paid exactly for the foundation that is required by the actual geotechnical conditions. Because these conditions cannot be accurately estimated until after the contract award, the payment mechanism is designed to allow the owner to accept the scope risk for the foundation while the design-builder carries the scope risk for the rest of the project. In this case, the design-builder will only need to select an appropriate mark-up for profit and overhead. As a result, the design-builder will probably declare a number that is lower than the one that would have been used in the first case because the owner has accepted this specific piece of the project's scope risk.

After award, both the owner and the design-builder must use the scope of work defined in the contract as the basis of work. Scope creep during design is an insidious thing that happens not because of the technical aspects of the project but, rather, because of the business psychology that is at work at this phase of the project. As already stated several times in this chapter, the design-builder's long-term business objective is to use this project as leverage to make it more competitive for future DB work with this and other owners. Thus, at early stages in the project, it will try to incorporate owner preferences and mitigate owner concerns about the quality of design submittals by not challenging review comments that do not appear to require major changes to the scope. However, the cumulative impact of many minor "improvements" can add up to be a major change in the final cost of the constructed product. Additionally, in the design phase they can also have a deleterious effect on design cost efficiency by unnecessarily creating unintended conflict with design in other disciplines. The overall effect is to increase cost and slow design progress, which will inevitably have an impact on project profitability.

The solution to scope creep during design is to sensitize both the design team and the design review team to its potential and to ensure that the design-builder's team is instructed to not incorporate seemingly innocuous minor changes into the final design until their effect on project budget and schedule has been evaluated.

Actions that Can Be Taken after Submittal Reviews

From the discussion in Chapter 1, the reader must remember that DB is generally all about the schedule. Most DB projects try to achieve an aggressive delivery schedule and, as a result, designers and designer reviewers do not usually have the luxury of arguing back and forth through the review comment process about how to optimize the final design. DB is definitely a delivery method where Thomas Paine's exhortation "Lead, follow, or get out of the way" should be a guiding theme. Design submittals are *always* on the critical path because they literally define what must be done in construction. Materials cannot be ordered, subcontractors cannot be scheduled, and construction cannot commence unless there is an approved design that details what will be built. Therefore, because haste is usually the enemy of quality in construction, the design QM system needs to have a well defined set of actions that are allowable after each design submittal review.

The major issue becomes defining the point at which the design for a given feature of work is released for construction. If the owner structures the design process around disciplinary work packages rather than the traditional percentage-based design submittals or the architectural system of schematic design, this decision becomes both easier and cleaner to make. For example, if a 30% design submittal for a building project is not approved on the initial submittal and the design-builder wants to begin the site work on the project, can that be done? Typically, a 30% design submittal will have 100% of the site work complete within it. So the decision will turn on whether or not the design is deficient in the site-work area.

However, if the same project were structured with a site civil design submittal, followed by a foundation design submittal, and so forth, the decision on whether to allow the earthwork on the site to commence becomes easy for both parties. If the site civil design package was not approved, the design-builder will not want to commence because the risk of potential rework due to design changes is too great. Additionally, this allows the owner's design reviewers to focus on discipline-specific design products, and this expedites the review process as well.

The next decision is whether to allow construction to commence on design elements that were found deficient, but where the redesign is relatively straightforward and understandable. Must the design-builder wait for re-review and approval? For instance, consider the case where the structural design of a bridge is disapproved because the engineer specified a bridge deck expansion joint system that is not permissible in the state standard specifications, which were incorporated by reference in the contract. The bridge subcontractor wants to order the precast concrete components upon which there were no review comments, so that they will be on-site in time to meet the schedule. Whereas in DBB a contractor can normally proceed at its own risk, some DB contracts require "released for construction" design documents before the owner will pay for the progress made in a given pay period. Therefore, in this example this shifts the decision out of the quality arena and into the financial area, as these components are very expensive. Thus, the design QM system could potentially affect the overall schedule. The obvious solution is to permit progress on those elements of design that were found to be compliant but, because this may entail a change to the specific contract terms, it should probably be addressed in the contract itself, permitting the owner to allow and pay for progress on a case-by-case basis. This flexibility actually strengthens the design QA process because the latter will no longer be viewed as an impediment to potential progress.

Without contract terms as described in the above paragraph, the same end result can be accomplished through the agreement reached during the initial design meeting. Thus, when conducting this initial meeting, it is important to keep the thought process on the project's life cycle rather than just focused on the design phase. It is also extremely important to document the output of that meeting for future reference, as situations arise that were not contemplated in either the contract or at the initial design meeting. Those two documents will form the foundation against which potential solutions can be measured to ensure that the spirit of both agreements is consistently applied throughout the course of the DB project.

Ability to Change Previously Approved Design Submittals

The final design quality issue revolves around those actions that will be viewed as permissible in changing previously approved design submittals. The owner's temptation is to mandate that once a design submittal is approved, it cannot be changed without owner concurrence. Superficially, this would seem to be logical because the owner often makes payments for design progress as design submittal milestones are satisfactorily met. Additionally, it seemingly protects the owner from the potential "bait and switch" concern that is always present in projects delivered using DB. Nevertheless, this forgets that the design process is inherently iterative and the achievement of a high-quality design necessitates coordinating and deconflicting various design disciplines as they are incorporated into the final construction documents. Thus, restricting the DOR and its subconsultants to having to consult with the owner's staff on every minor dimensional and/or material change creates a bias toward not making any changes for fear of delaying the schedule. This actually runs counter to good engineering practice as well as prevents the designers from enhancing the design.

Owners with more DB experience have found the above-described process to be overly legalistic and tending to strain rather than enhance the relations between the reviewers and the designers. A more enlightened approach is to require the DOR to submit a list of the deviations from the previously approved submittal. This method has several benefits to commend its use. First, it retains the "open books" style of design information flow by forcing the design-builder's design team to track changes in the design as they occur so they can be communicated to the owner in the next design submittal. This fosters rather than strains the mutual trust that is so essential to DB contracting. Next, it gives the DOR and its subconsultants a free hand to make the design as good as possible without imposing an onerous administrative requirement to seek external permission to make a change that everyone knows must be made anyway. This emphasizes the fact that design liability remains with the design-builder, who is held responsible for attaining a high standard of design quality. Finally, it leaves the door open for last-minute enhancements in a manner that will not adversely affect the project's schedule. This approach transforms the design QM system from the DBB "catch and punish" mode into one where innovation and creativity in design can flourish without constraint from the design QM process itself.

Design Quality Management Planning

Just as design details define construction quality requirements, it would follow that owners who must commit themselves to the cost of construction prior approving the project's final design—as happens in DB—would devote a significant portion of their DB solicitation packages to defining the required design QM process. That, in turn, would cause design-builders to prepare design QM plans that detail their proposed process for each specific project that can be evaluated as a part of the selection process. Unfortunately, in practice this is not occurring. A previous study of design QM requirements in 75 DB projects across the nation found that only 18% of the DB solicitation documents required a design QM plan to be submitted as part of the DB proposal (E. DeCorso, 2004, unpublished honors thesis, University of Oklahoma, Norman, Okla.: "How Owners Communicate Quality in Public Sector Design-Build Requests for Proposals"). Additionally, only 17 of those projects required a design QC plan post-award and only 2 took the next step by requiring a complementary design QA plan. Thus, the literature shows that design QM is an area where the greatest potential for improvement is present. Perhaps this is due to a lack of policy guidance in the area of design management due to the DBB practice of public engineering agencies traditionally doing much of their own design work, using in-house professional engineers. Thus, owners are not availing themselves of the opportunity to evaluate different design-builders' approaches to ensuring design quality by not asking for design QM plans prior to award. Owners give up control of the details of design by selecting DB project delivery, thus depending merely on the qualifications process to guarantee design quality; as in a pure design contract, this may not be sufficient. With the dominant organizational type being a constructor-led DB team (Songer and Molenaar 1996), the designer's client is no longer the owner. Therefore, it would seem to be imperative that the DB teams' approach to producing a quality design be evaluated prior to award. Thus, having a clear definition of design QM is imperative.

The Minnesota Department of Transportation (MnDOT) furnished an excellent definition of QM during the design phase of a DB project when it laid out the objectives of the Design Quality Management Plan as follows:

> The Design Quality Management Plan is intended to:
>
> - Place the primary responsibility for design quality on the design-builder and its designer(s).
> - Facilitate early construction by the design-builder.
> - Allow the Department to fulfill its responsibilities of exercising due diligence in overseeing the design process and design products while not relieving the design-builder from its obligation to comply with the contract. (Gonderinger 2001)

One can see that MnDOT's three-pronged approach not only satisfies its obligations for project oversight due to federal funding, but also ensures that the responsibility for the quality of the design is placed clearly on the design-builder's shoulders. It also speaks to achieving a major benefit accrued by the owner when selecting DB project delivery: project schedule compression through overlapping design and construction activities. Thus, it becomes important to not only adopt a good definition for design QM but to also clearly define the allocation of responsibilities between the DOT and the design-builder after project award.

In 2007 the Virginia DOT invested in the development of an updated DB QM plan template (VDOT 2007). This document furnishes an excellent generic outline for the development of a post-award DB QM plan for any type of project. VDOT supports the previous assertion that quality can be more easily managed if the design itself is broken down by discipline. The VDOT guide states: "The Quality Assurance and Quality Control procedures shall be organized by each type of engineering discipline (such as structural, civil and utilities)." The guide covers the entire procurement, design, and construction phases of a DB project. As such, the following discussion will be restricted to the design portion; the construction planning will be covered in Chapter 6, Construction Quality Management in the Design-Build Project.

General Design Quality Requirements

The design QM plan must first ensure that its users understand to what processes it will apply. The VDOT guide begins with the following set of instructions:

> All QA and QC procedures proposed by the Design-Builder for the design process shall be included in the QA/QC plan. . . . These [design QM] procedures shall specify measures to be taken by the Design-Builder
>
> > (1) to ensure that appropriate quality standards are specified and included in the drawings, specifications, and other design submittals and to control deviations from such standards, it being understood and agreed that no deviations from such standards shall be made unless they have been previously approved by the Department at the Department's sole discretion, and (2) for the selection of suitability of materials, and elements of the Work that are included in the Project. (VDOT 2007)

By using this verbiage the owner is effectively making the details of this plan the contract requirement for measuring the effectiveness of the design quality process itself. As previously mentioned, the owner can choose to assign the majority of the QM responsibilities to the design-builder and, in doing so, to relegate itself to quality oversight or quality audits as its only enforcement mechanism. By requiring the development of a detailed design QM plan like the VDOT guide, it establishes the baseline against which it can conduct subsequent quality audits as the project progresses.

Next, it is important that the design QM plan details the roles and responsibilities of all the players in the design QM process. Virginia requires that the design QM plan "include an organizational chart of who will perform each task and their firm, discipline, name, qualifications, duties, responsibilities and authorities. All subcontractors shall be included in the plan" (VDOT 2007). Just as importantly, the plan needs to specifically reference those areas of the project scope of work that will require greater attention to detail and care during the design process to ensure that both the owner and the design-builder are aware of those areas where the designers will be challenged by the technology and which, as a result, will need extra attention during the design QM process. VDOT does this as follows:

> In addition, the design QA/QC plan should identify those elements of the contract, drawings, specifications, and other design submittals, if any, requiring special construction QA and/or QC attention or emphasis, including applicable standards of quality or practice to be met, level of completeness and/or extent of detailing required, or Special Provisions to the Road and Bridge Specifications. (VDOT 2007)

Design Quality Control Activities Design QC is the first line of defense in the design QM process. Design QC is performed by the designers themselves. These activities can range from the simple manual spot-checking of computer calculations to the review of drawings to ensure that proper line weights are used. The *Transportation Research Circular E-C074* (TRB 2005) furnishes elegant definitions for the difference between QC activities and QA activities. It states that QC involves "doing things right," whereas QA involves "doing the right things." As such, the design QM plan must have clear definitions of the activities that fall in the design QC realm. The VDOT guide does this as follows:

> The Design QC plan shall include the level, frequency and methods of the procedures or actions necessary to accomplish the following tasks:
>
> 1. Preparing all drawings.
> 2. Checking math and engineering computations and other aspects of drawings.
> 3. Reviewing of the technical accuracy of the plans.
> 4. Reviewing the specifications.
> 5. Ensuring that the plans conform to contract requirements.
> 6. Checking the form, content and spelling in plans.
> 7. Reviewing coordination with other design disciplines.
> 8. Reviewing designs for omissions.
> 9. Reviewing the sequence of construction.
> 10. Ensuring there will be no utility conflicts.
> 11. Constructability reviews. (VDOT 2007)

Design Quality Assurance Activities When determining what goes into the design QA portion of the plan, the owner must determine just how deeply involved it will be in the QA process throughout the project. Once again, the Virginia guide provides a useful checklist for this process.

> The Design QA plan shall include the level, frequency and methods of the procedures or actions necessary to accomplish the following tasks:
>
> 1. Evaluating whether the designer addressed problem appropriately.
> 2. Evaluating if correct analyses were applied.
> 3. Evaluating whether qualified, independent personnel were assigned to the QC tasks and stamped the drawings.
> 4. Evaluating whether solutions are practical and cost-effective.
> 5. Evaluating whether the design is within appropriate range based on past experience.
> 6. Levels of DOT review and/or approval.
> 7. Final, independent review. (VDOT 2007)

From the above discussion, it can be concluded that owner DB policy documents such as guidelines and model RFPs should contain specific guidance with regard to the owner's design QM approach. They should also contain guidance on how the owner's design QM program can be modified and adjusted to fit the specific needs of each project. Creating a specific policy requirement to address the required quality activities at this fundamental policy level articulates the importance to the owner's project managers of considering project design quality from the very inception of the project. It also creates a mechanism to map the conceptual design quality requirements identified in the early stages of project development directly into the DB project procurement documents. This will not only furnish much-needed guidance to the owner's project managers but also promote consistency in a given owner's DB projects, making estimating the cost of these activities easier for the competing design-builders.

Design-Build Design Submittal Process

One of the traditional ways that owners have ensured quality design is by being able to fully review the design before it is advertised for bids. In DB, owners do not have this same opportunity. One of the major advantages of DB is schedule compression, which happens by being able to start construction before the full design is finalized. In fact, in a recent research project on QM in DB for transportation projects, 85% of surveyed state DOT respondents (hereafter referred to as "our DB QA study") (Gransberg et al. 2008) indicated this as a reason for implementing DB. Another advantage of DB is the transfer of risk from the owner to the design-builder, and that study found that 53% of the state DOT respondents also indicated this as a reason for implementing DB. In a DB contract, the design-builder is responsible for the adequacy of the design in relation to the contract documents. Owners must be aware that "increased control over project design might not only reduce potential design-build benefits but might also carry with it the risk of liability for the entire project" (Wichern 2004). Arkansas clearly states this in its *Design-Build Guidelines and Procedures*:

> With Design-build contracting, the design risk is placed with the Design-build firm, and the Department's review will determine if the proposed design meets the objectives of the Contract Provisions. (AHTD 2006)

Thus, many owners that do place the responsibility for design QA and QC on the design-builder use specified design review checkpoints—a design PQA activity—to ensure that the design is proceeding according to contract requirements. This also fulfills the owner's responsibility to the public to deliver projects that have been designed and built in accordance with public law and good engineering practice. These checkpoints exist so that the design-builder's final design is acceptable to the owner and is in accordance with the performance criteria contained in the contract documents.

Design Review Checkpoints

The RFP content analysis found in our DB QA study that there are two general ways in which design review checkpoints are determined. These are summarized in

Table 4-4. The first method, defined reviews, specifically articulates them in the RFP. The owner specifies in the RFP which reviews it will conduct and what must be included in the reviews. The design-builder must then account for the required reviews in its proposal and in the project schedule.

The following example comes from a Mississippi DOT (MDOT) RFP and outlines the design review requirements for a project:

> Preliminary Design Phase (Minimum 30% Plans): The *Contractor* will prepare and submit a single preliminary design submittal for the entire project. . . . Final Design Review Phase (100% Plans): Final Design may be broken down into packages (i.e., Roadway, portions of Bridges, Drainage, etc.) as determined by the *Contractor*. Following completion of the design for each submittal for the Project, the *Contractor* shall prepare and submit a Final Design Submittal for review by MDOT. . . . Released for Construction Documents: Following the incorporation of MDOT's comments from the Final Design Review Phase, the *Contractor* shall prepare and submit a Release for Construction submittal to MDOT for MDOT's final review and Released for Construction stamp. (MDOT 2005; emphasis added)

This is by far the most common way to identify the required reviews. In our DB QA study's solicitation document analysis, 41 projects had design reviews as a requirement of the contract. Of these 41 projects, 83% told the design-builder at what point the design would be reviewed.

The second approach, proposed reviews, is to allow the design-builder to propose the schedule of design reviews in their response to the RFP or during negotiations after the award of the contract. This is the stated policy of the Arkansas DOT: "There will be no pre-defined reviews scheduled by the Department. The selected firm and the Department will decide on the appropriate timing of reviews

Table 4-4 General Design-Build Design Review Categories

Type	Design-Builder Responsibility	Owner Responsibility	Comments	Percent of Projects in DB QA Study Content Analysis
Defined Reviews	To be responsive, must follow defined reviews in contract documents	Defines reviews in the RFP	Reviews may be performed by design-builder, DOT, or 3rd party	83%
Proposed Reviews	Propose design reviews for project as part of proposal or after award of contract	Accepts or rejects proposed design reviews	Reviews may be performed by design-builder, DOT, or 3rd party	17%

Source: Gransberg et al. (2008), Table 10, with permission from the Transportation Research Board of the National Academies.

during execution of the contract" (AHTD 2006). The Washington State DOT (WSDOT) used this approach for its Thurston Way Interchange project:

> For any designs for which early construction reviews will not be conducted, at least one design review shall be conducted before completion of 100 percent design. The percentage of design will be mutually agreed upon between the *Design-Builder* and WSDOT, but should be near the mid-point of design. (WSDOT 2000; emphasis added)

In our DB QA study's solicitation document content analysis, DOTs employed this approach in only 17% of the projects reviewed.

Appropriate Number of Design Reviews

In addition to how the design reviews are defined, the number of required design reviews by the owners varies across the nation. However, our DB QA study's content analysis identified three main approaches:

- No formal review prior to final (release-for-construction) design review.
- One review prior to the final design being released for construction.
- Multiple reviews prior to the final design review.

Also, in many instances the design-builder is encouraged to request informal reviews that are not required but allow the owner to provide more frequent input to ensure that the final design will meet the contract requirements. These reviews are often called "over-the-shoulder" or "oversight" reviews to indicate that the design process will not stop proceeding to wait for comments that result from these informal reviews. Table 4-5 provides a summary of the different categories of the

Table 4-5 Required Number of Design Reviews

	Percent of Projects in DB QA Study Content Analysis	Comments
No review prior to final	15%	Owner still provides oversight and comments informally
One review prior to final	56%	Can be anywhere from preliminary design until just before the final design review
Multiple reviews prior to final	29%	The exact number of reviews can range from one to two for every major feature of work

Source: Gransberg et al. (2008), Table 11, with permission from the Transportation Research Board of the National Academies.

required number of design reviews and the corresponding percentage of occurrences in our RFP analysis.

No Mandated Reviews When there is no owner-mandated design review checkpoint required before the final design is released for construction, the burden of design compliance is fully placed on the design-builder. In theory, this is one of the benefits of utilizing DB project delivery. However, the owner must still provide assurance that the contract will be completed with all the requirements met in a timely manner. In the RFPs analyzed in our DB QA study, 41 mentioned the design review requirements. 15% used the approach of no owner-mandated design review checkpoints before the release-for-construction design review. The Minnesota DOT detailed its design PQA approach in one RFP as follows:

> The Department will participate in oversight reviews and reviews of early construction as part of its due diligence responsibilities. If the Department, in its review, observes that the Design-Builder is not complying with contract requirements and/or that the QC/QA checks are not complete, it will notify the Design-Builder in writing that construction may not proceed until the noted items are corrected. The Department's oversight review and comments will not constitute approval or acceptance of the design or subsequent construction. (MnDOT 2001)

This PQA activity (sometimes termed "due diligence") must be accomplished through an oversight approach, as stated in the MnDOT RFP referenced above, or by an audit approach referenced in the Utah RFP below, which describes the design-builder's review procedures. The Utah RFP also shows that although the owner does not conduct a progress design review, the design-builder must do so with oversight from the owner:

> The *Design-Builder* will review all designs to ensure the development of the plans and specifications are in accordance with the requirements of the Contract. . . . The Department will audit, as needed, the *Design-Builder* processes and Design Documents to verify compliance with the Contract Documents. The Department will be invited to attend all reviews. . . . The *Design-Builder* shall conduct oversight reviews, and the Department may participate in these reviews and comment as requested or as it otherwise deems necessary. . . . The *Design-Builder* shall determine the materials to be compiled for each review. Formal assembly and submittal of drawings or other documents will not be required, but the Design-Builder is encouraged to provide informal submittals to facilitate reviews. The review may be of progress prints, computer images, draft documents, working calculations, draft specifications or reports, or other design documents. . . . The *Design-Builder* will conduct informal milestone reviews at approximately the 60% stage of project elements to determine whether the Contract requirements and design are being

> followed. The Department will be invited to attend these reviews. (UDOT 2005; emphasis added)

The Utah RFP goes on to discuss the design review process for the final design deliverable.

> When the *Designer* has completed a design package to 100%, and the package has been checked and audited, a formal design submittal is assembled and distributed for review, including plan sheets, calculations, specifications, and other pertinent data. The Designer shall prepare for these reviews a full set of drawings and other documents stamped "Checked and Ready for Review." . . . After the 100% comments have been addressed and the design documents have been checked and audited, a "ready to be released for construction" submittal package is assembled and distributed to the *Design-Builder* and the Department for release for construction. (UDOT 2005; emphasis added)

To preserve the definition of design liability, Utah also requires the design-builder to complete a certification process on the final design package, and specifies the time limit to which the owner must adhere to furnish timely acceptance:

> When a design package is ready to be released for construction, the *Design-Builder* shall certify all of the following related to the Work:
>
> - The design is in accordance with the Contract requirements.
> - The design has been checked in accordance with UDOT accepted quality procedures.
> - No design exceptions exist that have not previously been approved by the Department.
> - The Department will conduct its review and accept or reject the final design package within seven (7) Working Days of receipt of the final design documents. (UDOT 2005; emphasis added)

Single Design Review The second category of DB design review is where the owner requires a single official review of the design before the review of the final design deliverable. This gives the owner an intermediate point at which to verify that the design development is proceeding in accordance with the contract requirements and is progressing according to the schedule. The Mississippi DOT uses this type of design review for their DB projects. An example is:

> The *Contractor* will prepare and submit a single preliminary design submittal for the entire project. Preliminary design shall include roadway plan and profile, bridge type, selection layout, drainage, erosion control, signing, architectural and traffic control plans. MDOT will review Preliminary Design Submittals within 21 Days of the submittal. (MDOT 2005; emphasis added)

MDOT also provides for an "optional design review" with the following RFP clause:

> At the request of the *Contractor*, MDOT will provide optional design reviews on design packages as requested by the *Contractor*. MDOT as appropriate will review optional design Submittals within 14 Days. (MDOT 2005; emphasis added)

This RFP goes on to define the final design review process as follows:

> Final Design may be broken down into packages (i.e., Roadway, portions of Bridges, Drainage, etc.) as determined by the *Contractor*. Following completion of the design for each submittal for the Project, the *Contractor* shall prepare and submit a Final Design submittal for review by MDOT. . . . Following the incorporation of MDOT's comments from the Final Design Review Phase, the *Contractor* shall prepare and submit a Release for Construction submittal to MDOT for MDOT's final review and Released for Construction stamp. (MDOT 2005; emphasis added)

Another example comes from the Washington State DOT in the RFP for its Thurston Way Interchange project. The exact point of the design review is not listed, but it is left to be decided upon execution of the contract:

> For any designs for which early construction reviews will not be conducted, at least one design review shall be conducted before completion of 100 percent design. The percentage of design will be mutually agreed upon between the *Design-Builder* and WSDOT, but should be near the mid-point of design. (WSDOT 2000; emphasis added)

The requirement of only one official review by the owner is, by far, the most popular design review process currently used, as found in our RFP analysis. 56% of the RFPs analyzed in our study used this type of design review process.

Multiple Design Reviews In the final category of design reviews, the owner requires more than one official owner review before the design can be released for construction. This was the process found in 29% of our DB QA study's RFPs that included information about design reviews. The Maine DOT required in one RFP that "formal design package submittals shall be made . . . at the 50% and 80% design development stage of any design package intended to be RFC [released-for-construction]" (MaineDOT 2003). The EFLHD also requires more than one design review before the design is released for construction. In an RFP they state the reasons for the reviews:

> Initial submittals are intended to provide the *Contractor* a means of proposing and obtaining acceptance for horizontal and vertical alignment deviations from the Government preliminary design plans; deviations from the

> Government preliminary bridge Type, Size, and Location (TS&L) plan; and changes in basic parameters of the project. . . . Intermediate Design Submittal. The purpose of this submittal is to ascertain that the design is progressing in accordance with the requirements of the project, that existing field conditions have been properly identified and dealt with, and that the *Contractor* has coordinated the design with EFLHD, NPS, the permitting agencies, and the utility companies. (EFLHD 2001; emphasis added)

We found two variations of this category that require mention in this section. The first is when the owner requires an independent design quality assurance firm to do the design reviews, with the owner only providing limited oversight. This was the situation with the S.H. 130 project in Texas. The Texas Turnpike Authority's RFP stated:

> DQAM [design quality assurance manager] will conduct a formal over-the-shoulder review presentation to the TTA [Texas Turnpike Authority] at the TTA's office. The over-the-shoulder review presentation will be held, following the DQAF's [design quality assurance firm's] approval of: the Corridor Structure Type Study Report; the Preliminary (30%) Design Submittal; the Intermediate (65%) Design Submittal; and the Final (100%) Design Submittal. . . . Developer's designer shall furnish to the DQAF at least five (5) mandatory design submittals, and if necessary, any resubmittals. (TTA 2001)

The second variation is when the owner requires certain design review, and attends the reviews, but is not the party responsible for the review. In the following example from Washington State, the DB firm was responsible for the formal design reviews with the DOT in attendance:

> The DQA [design quality assurance] Manager will conduct formal milestone reviews at the 30%, 60%, and 90% (or as otherwise agreed by the WSDOT and Design-Builder) stage of project elements to determine whether the Contract requirements and design are being followed and that QC/QA activities are following the approved QMP. . . . The DQA Manager shall compile and maintain documentation of the review. The Department will be invited to attend these reviews. (WSDOT 2004)

In the vein of deciding the appropriate number of owner design reviews for a given project, it is interesting to note that the U.S. Army Corps of Engineers (USACE) changed its DB design reviews policy in 2007, reducing the number of reviews from four (30%, 60%, 90%, and final) to two (intermediate and final) (USACE 2006). The reason for the change was to reduce the potential for delays due to waiting for government reviews. In a personal communication with the author, Joel Hoffman of USACE explained the rationale as: "The philosophy is that once the designer of record approves construction and extension of design submittals, the builder can proceed—don't wait on us, unless there is a specific government approval required." Thus, one critical issue in determining the appropriate number of design reviews is the need for the design-builder to maintain an aggressive schedule. If the project is not schedule-constrained, the DB design reviews can afford to inject more design review points. Conversely, design reviews can be minimized on a fast-track project.

Over-the-Shoulder Reviews In addition to the design reviews outlined above, another notable trend is the inclusion in the RFP of a statement inviting the design-builder to request informal over-the-shoulder reviews to ensure that the design is progressing according to the contract requirements without the need to prepare a specific design submittal package, and to provide owner input to the design where it will be both desired and helpful. These reviews fall into the owner PQA category. These statements are included in RFPs regardless of the number of required design reviews. Almost always, however, a statement is also included that removes liability from the owner for any comments that may be incorporated into the design from the informal reviews. The following extract comes from the EFLHD RFP referenced above:

> Over-the-Shoulder reviews may be scheduled by the Contractor or EFLHD. Over-the-Shoulder reviews are strongly encouraged to enhance the partnering efforts between the Contractor and the Government. . . . The number and timing of the reviews will be discussed at the Start-up Conference. . . . ver-the-Shoulder reviews will be conducted for informal review of designs. The intent of Over-the-Shoulder reviews is to provide guidance to the Contractor during the course of the project. Over-the-Shoulder reviews do not take the place of the Overall Project Submittals. (EFLHD 2001)

The Washington State DOT included this in one RFP:

> Throughout the design process, the Design-Builder may request additional oversight visits by WSDOT to discuss and verify design progress and to assist the Design-Builder and/or its designer(s) in resolving design questions and issues. (WSDOT 2000)

Design reviews are an integral part of any design QA program. They ensure the constructability of the project as well as that the design meets the contract requirements. Even though the design-builder is responsible for both of these elements in DB, the owners must assure itself that the design-builder is carrying out its responsibility. This is done by owner design reviews using one of the three approaches outlined above.

Design Review Responsibility

Communicating who is responsible for the design reviews is also essential to the smooth execution of these quality activities. This can be done in a variety of ways, including lists, charts, diagrams, or designating responsibility in contract clauses. Table 4-6 is an example from a Louisiana DOT DB RFP that provides a good example of how to effectively communicate design review responsibility.

In addition to deciding which reviews will be conducted and when, deciding who will perform the reviews is just as critical to the success of DB projects. Since the owner is not performing the design with its own designers, design QA and QC responsibilities will shift in DB. One can see in Table 4-6 that the Louisiana DOT has assigned virtually all the design QA and QC responsibility to its design-builder, only entering the process to verify the as-built plans. Remembering that the design phase defines the standard of quality for the constructed project, it is imperative that the design documents are professionally reviewed and checked to ensure a quality project. Our DB QA study sought to identify the trend in design quality responsibilities by asking the respondents to indicate which entity was primarily assigned the responsibility for a list of common design QM tasks. However, a large number of the respondents did not confine themselves to furnishing a single answer to each question. Many indicated that the responsibility for the tasks was indeed shared among some combination of the agency, the design-builder, and the agency's consultants. This response yielded valuable information regarding the distribution of design QM responsibility among the parties to a DB contract. Table 4-7 summarizes the survey responses to the question of assigned responsibility for design QM tasks.

Table 4-7 shows that the design-builder or a third-party consultant is often given more responsibility than is seen in traditional DBB in performing design

Table 4-6 Communicating Design Review Responsibilities: Louisiana Department of Transportation and Development

Stage of Design Development	Design Check and Certification to Design-Builder	Design Review
Definitive Design	Designer and DQCM	DQCM
Interim Review	Designer and DQCM	DQCM
Readiness for Construction Design	Designer and DQCM	DQCM
Final Design	Designer and DQCM	DQCM
Working Plans and Related Documents	Designer and DQCM	DQCM
As-Built Plans	Designer and DQCM	LADOTD's designated representative
Major Temporary Components	Designer and DQCM	DQCM
Temporary Components	Designer and Checker	Not applicable

DQCM, design quality control manager; LADOTD, Louisiana Department of Transportation and Development.

Source: LADOTD (2005).

Table 4-7 Survey Responses for Design Quality Management Task Responsibility

Who performs the following design quality management tasks?	Agency Personnel	Agency-Hired Consultant	Design-Builder's Design Staff	Design-Builder's Construction Staff
Checking of design calculations (QC)	15.4%	15.2%	68.7%	0.8%
Checking of quantities (QC)	13.8%	11.2%	53.1%	21.8%
Review of specifications (QC)	32.9%	25.0%	38.9%	3.2%
Technical review of design deliverables (QC)	30.9%	28.6%	40.0%	0.6%
Acceptance of design deliverables (QA)	57.9%	15.8%	22.9%	3.3%
Approval of final construction plans and other design documents (QA)	82.0%	5.2%	9.9%	2.9%
Approval of progress payments for design progress (QA)	81.8%	9.1%	2.0%	7.1%
Approval of post-award design QM/QA/QC plans (QA)	84.4%	9.7%	5.9%	0.0%

Letters in parentheses indicate type of task: QC, quality control; QA, quality assurance, QM, quality management.
Source: Gransberg et al. (2008), Table 13, with permission from the Transportation Research Board of the National Academies.

QM tasks that lead up to the final acceptance of the design. Public agencies are assigning the design-builder the responsibility for design QC tasks, such as checking design calculations, checking quantities, technical review of design deliverables, and review of specifications. Because these tasks are primarily associated with the production of design deliverables, the owner is facilitating the overall schedule by stepping back from these tasks and giving the design-builder control. Additionally, it effectively prevents the owner's unintentional assumption of design liability through directive design review comments (Gransberg et al. 2006). When it comes to the design QA tasks of accepting and approving *final* construction plans and design documents, owners have, by and large, retained the responsibility. This makes sense since owners still have ultimate responsibility for the design, construction, and final quality of each project. Public owners cannot contractually assign their public duty to another party. They can have design-builders and third-party consultants help achieve an assurance of quality, but at the end of the day they must be able to affirm that each project has been constructed to the requisite quality level.

Our DB QA study asked each respondent to cite the number of DB projects in which its agency had been involved. This allowed the responses to be divided by experience level. The responses from agencies with more than five DB projects were assembled as a single group to compare to the responses of the total population. The idea was to capture the potential differences between agencies whose QM system had been able to benefit from lessons learned in early DB projects and those agencies that were embarking on their first series of DB projects. Intuitively, those with more DB QM experience should have had a better knowledge of how to distribute the responsibility for design quality. Our study found that in every category the more experienced owners retained a higher percentage of responsibility for design QC tasks, and most of the time they retained responsibility for design QA tasks and gave less responsibility to the design-builder than did the general population. The differences between the experienced respondents were especially significant in the categories of technical review of design deliverables, acceptance of design deliverables, review of specifications, approval of progress payments for design, and the post-award design QM plan approval. These are the points in the design process where the final design decisions are made, and the experienced owners seemed to feel the need to impose themselves in the design QM process at these points.

Additionally, only 20% of the more experienced owners indicated that they were using a third-party consultant to perform design QA tasks. This was less often than the less experienced owners, where a consultant was used by 50% of the respondents. This makes sense because an owner with little or no previous DB experience could mitigate the risks associated with inexperienced personnel by retaining an experienced DB consultant to assist it with the QM during its first series of DB projects.

References

Arizona Department of Transportation (ADOT). (2001). "Design-build procurement and administration guide." *Arizona Department of Transportation, Intermodal Transportation Division,*<www.azdot.gov/Highways/ConstGrp/Value_Quality/PDF/DesignBuildGuide.pdf> [March 3, 2010].

Arkansas State Highway and Transportation Department (AHTD). (2006). *Design-build guidelines and procedures.* Arkansas State Highway and Transportation Department, Little Rock, Ark.

Beard, J. L., Loulakis, M. C., and Wundram, E. C. (2001). *Design-build planning through development.* McGraw-Hill, New York.

Burati, J. L., Weed, R. M., Hughes, C. S., and Hill, H. S. (2003). "Optimal procedures for quality assurance specifications." *Federal Highway Administration Report No. FHWA-RD-02-095,* U.S. Dept. of Transportation, Federal Highway Administration, Washington, D.C.

Design-Build Institute of America (DBIA). (1998) *DBIA manual of policy statements,* Design-Build Institute of America, Washington, D.C.

Ernzen, J., and Feeney, T. (2002). "Contractor-led quality control and quality assurance plus design-build: Who is watching the quality?" *Transportation Research Record* 1813, 253–259.

Eastern Federal Lands Highway Division (EFLHD). (2001). *Request for Proposals IBC-8888(012),* Book 1, Federal-Aid Project No. IBC-8888(012), District of Columbia Department of Public Works, Washington, D.C.

Federal Highway Administration (FHWA). (2006). "Design-build effectiveness study: Final Report to Congress as Required by TEA-21." *USDOT Federal Highway Administration,* <http://www.fhwa.dot.gov/reports/designbuild/designbuild0.htm> [Aug. 30, 2006].

FHWA. (1998). "Life-cycle cost analysis in pavement design: In search of better investment decisions." *Publication No. FHWA-SA-98-079,* U.S. Dept. of Transportation, Federal Highway Administration, Washington, D.C.

Gharaibeh, N. G., Button, J. W., and Jalvi, P. (2005) *Pavement aspects in design-build contracting for highway projects synthesis of highway practice,* Report No. TTI-2005-1, Texas Transportation Institute, Texas Department of Transportation, Austin.

Gonderinger, C. (2001). "TH 14/218 design-build project technical memorandum quality management plan (QC/QA requirements)." *Technical memorandum,* Minnesota Department of Transportation, St. Paul, Minn.

Gransberg, D. D., Molenaar, K. R., and Datin, J. N. (2008). "Quality assurance in design-build projects." *NCHRP Synthesis 376,* Transportation Research Board of the National Academies, Washington, D.C.

Gransberg, D. D., and Windel, E. (2008). "Communicating design quality requirements for public sector design/build projects." *ASCE J. Mgmt. Eng.,* 24(2), 105–110.

Gransberg, D. D, Koch, J. E., and Molenaar, K. R. (2006) *Preparing for design-build projects: A primer for owners, engineers, and contractors,* ASCE Press, Reston, Va.

Gransberg, D. D., and Molenaar, K. R. (2004). "Analysis of owner's design and construction quality management approaches in design-build projects" *ASCE J. Mgmt. Eng.,* 20(4), 162–169.

Kirkpatrick, E. G. (1970). *Quality control for managers and engineers.* John Wiley & Sons, Inc., New York.

Louisiana Department of Transportation and Development (LADOTD). (2005). *Request for proposals, new Mississippi river bridge: St. Francisville to New Roads, including approaches and connecting roadways Route LA 10, design-build project,* State Project No. 052-02-0024, Louisiana Department of Transportation and Development, Baton Rouge, La.

Low, S. P., and Yeap, L. (2001). "Quality function deployment in design/build projects." *Journal of Architectural Engineering,* 7(2), 30–39.

Maine Department of Transportation, (MaineDOT). (2003). *Request for proposals, I-295 Commercial Street connector,* FHWA Project No. IMD-7589(300); MaineDOT Project No. 7589.30, Maine Department of Transportation, Augusta, Maine.

Minnesota Department of Transportation (MnDOT). (2005). *Addendum 5 project management book 2B, Trunk Highway 52 Oronoco design-build project,* Minnesota Department of Transportation, St. Paul, Minn.

MnDOT. (2001). *Part I: Scope of work T.H. 100–Duluth St. S.P. 2735-172, design-build request for proposals,* Minnesota Department of Transportation, St. Paul, Minn.

Mississippi Department of Transportation (MDOT). (2005). *Request for proposals, addendum 1, a design-build project bridge replacement on US 90 over St. Louis Bay Hancock and Harrison Counties, Mississippi,* Project No. ER/BR-0003-01(098) 104555/301000–US 90 St. Louis Bay Bridge Replacement, Mississippi Department of Transportation, Jackson, Miss.

National Cooperative Highway Research Program (NCHRP). (1979). *NCHRP synthesis of highway practice 65: Quality assurance,* Transportation Research Board of the National Academies, Washington, D.C.

New Mexico State Highway and Transportation Department (NMDOT). (2001) *Request for proposals, US 70 Hondo Valley,* AC-MIP-070-41(35) 264, New Mexico State Highway and Transportation Department, Santa Fe, N.M.

New York State Department of Transportation (NYSDOT). (2005). *Design-build procedures manual,* Vol. 1, New York State Department of Transportation, Albany, N.Y.

Pantazides, L., (2005). "Managing quality on transportation mega projects." *Proc., ASQ World Conference on Quality and Improvement,* Vol. 59, American Society for Quality (ASQ), Milwaukee, Wisc., 289–297.

Pheng L. W., and Yeap, L. (2001). "Quality function deployment in design/build projects." *ASCE J. Arch. Eng.* 7(2), 30–40.

Postma, S. E., Cisneros, R., Roberts, J., Wilkison, R., Clevenger, J., and Eastwood, A. (2002). "I-15 corridor reconstruction project design/build evaluation." *2001 annual report: Report No. UT-02.11.* Utah Department of Transportation, Research Division, Salt Lake City, Utah.

Potter, J. M., and McMahon, D. K. (2006). *Selecting a quality control/quality assurance program for a mega design-build project,* Rummel, Klepper, and Kahl, LLP, Baltimore, Md.

Songer, A. D., and Molenaar, K. R. (1996). "Selecting design-build: Private and public sector owner attitudes." *ASCE J. Eng. Mgmt,* 12(6), 47–53.

Stefani, A. M. (2004). *Managing risk in the Federal-Aid Highway Program,* Federal Highway Administration Report MH-2005-012, Government Printing Office, Washington, D.C.

Tam; W. O., Bowen, J. P., Smith, K. A., and Weight, T. J. (2003). "State highway 130 exclusive development agreement construction quality assurance program." *Asphalt Paving Tech.,* 72, 656–665.

Texas Turnpike Authority (TTA). (2001). *Request for proposals to construct, maintain and repair the SH 130 turnpike through an exclusive development agreement,* Texas Turnpike Authority, Texas Department of Transportation, Austin, Tex.

Transportation Research Board (TRB). (2006). *Design-build: A quality process: Transportation Research Circular E-C090,* Transportation Research Board of the National Academies, Washington, D.C.

TRB. (2005). *Glossary of highway quality assurance terms: Transportation Research Circular E-C074,* Transportation Research Board of the National Academies, Washington, D.C.

Tyborowski, T., Primack, M., and Matthews, K. (1997). "Privatization: A closer look." *1997 AACE Int. Trans.,* AACE International, Morgantown, W.V., CC.03.1–CC.03.5.

Utah Department of Transportation (UDOT). (2005). *Request for proposals, SR-92; east of I-15 in Lehi, Utah County,* Project No. STP-0092(5)1, Utah Department of Transportation, Orem, Utah.

U.S. Army Corps of Engineers (USACE). (2006). "MILCON transformation model RFP." *Department of the Army, U.S. Army Corps of Engineers,* <http://www.wbdg.org/ccb/ARMYCOE/COEECB/ecb_2006_13.pdf> [March 10, 2010].

Virginia Department of Transportation (VDOT). (2007). *Quality control, quality assurance, independent assurance and independent verification guide for use on public-private transportation act & design build projects,* Virginia Department of Transportation, Richmond, Va.

Washington State Department of Transportation (WSDOT). (2004). *Request for proposals, Everett HOV design-build project,* Washington State Department of Transportation, Olympia, Wash.

WSDOT. (2000). *Request for proposals, Thurston Way interchange, SR500 MP3.15 to MP 4.73, Clark County,* Washington State Department of Transportation, Olympia, Wash.

Wichern, S. (2004). "Protecting design-build owners through design liability coverage, independent construction managers, and quality control procedures." *Transportation Law Journal,* 32(1), 35–56.

FIVE

Design Administration Case Studies

This chapter synthesizes some of the fundamental issues from previous chapters through the presentation, analysis, and discussion of five case studies. The cases are drawn from real projects and have been sanitized to prevent identification of the actual participants. The format for each case covers the following: first, the facts and situation; second, a discussion of the issues and positions; third, the outcome; and fourth, lessons learned.

Case 5-1: "When Is Design *Really* Design?"

A public owner advertised a DB for a major bridge using the two-step DB method with a firm, fixed price. As part of the RFQ, the owner stated that it would only consider two possible types of bridges. Each DB team was instructed to indicate which type of bridge it intended to design and build, to permit the owner to evaluate their qualifications in light of the different technical challenges associated with each bridge type. The RFQ clearly stated that both types were considered equal and that the bridge type selection would not influence the ultimate award. As a result of the RFQ requirements, one of the competing DB teams felt the need to do "quick and dirty" conceptual designs for each bridge type before they made their decision because they suspected that one type would cost less. They also believed that the owner's budget was going to be tight and that the key to winning this project would be submitting the lowest proposed price.

During preparation of the qualifications submittal, the designer-of-record (DOR) for the team completed the conceptual designs and confirmed that one type was significantly less costly than the other. Therefore, the team selected the low-cost option as the one they would propose in Phase 1. This team was successful, making the short list and then winning the project with the lowest price. In fact, their price was the only one that fell within the owner's budget.

During an internal design review at a point where the major elements of the main span were nearly completely designed, the constructor's estimator reported that the quantities of work for both concrete reinforcing and structural steel had grown to a point where they were about 15% greater than the quantities that were used in the price proposal. This scope creep was going to seriously threaten the target profit on the job. The DOR was directed to revisit the design and determine why there was a discrepancy in the quantities. Upon investigation, it was found that the structural engineer who had completed the conceptual design during the Phase 1 competition had used the wrong seismic loads in that design, and that increased quantities of steel were required to meet the appropriate seismic code. As a result, the steel quantities could not be reduced. The structural engineer became personally aware of the mistake shortly after contract award when he was directed to complete the design, but had not informed his superior of the problem. The cost of the change in steel quantities ran into the millions of dollars.

The design subcontract under which the DOR was retained utilized a disputes resolution clause that required mediation before litigation could be undertaken. During the mediation, the engineer who made the mistake took the position that the conceptual design he completed to allow the DB team to make their bridge type selection was not a "true design" for which he should be held liable. He cited the following reasons for this defense:

- There was no assurance that the team would make the short list or win the project. Therefore, the conceptual designs were done merely to define an "order of magnitude" for each option and, as a result, the design-builder had no right to rely on their technical accuracy.
- He had not produced sealed construction drawings or specifications. Thus, the design-builder should have known that the design would change as all designs do in normal design practice.
- The lack of a seal indicated that the conceptual designs were a work-in-progress and a design professional is not liable for the result of his or her design errors until the works containing the errors are formally sealed. In other words, the designer is not done designing until the seal is on the design itself and thus has the right or obligation to correct his or her errors as soon as they are identified. Hence, the final design is evidence that the engineer found his earlier error and corrected it.

Issues and Discussion

The issues in this case focus on the roles and responsibilities of the DOR for design errors or omissions made before DB contract award, as well as the disposition of errors identified after contract award. Specifically in this case, the mediator had to seek to both answer and find solutions for the following questions:

- Is a mistake or oversight made during proposal preparation and before contract award of the same nature as a design error or omission

committed after contract award and hence compensable under the designer's professional liability policy?
- Does a design product produced by a design professional under contract to furnish design services require a seal to make it "official"?
- Did the design-builder have a right to rely on the conceptual designs produced in response to the RFQ?
- Who should be responsible for the cost overrun as a result of the design error, assuming it truly was a design error?

Outcome

During mediation, the engineer freely admitted that he made a mistake during the Phase 1 conceptual design. He stated that because there was no assurance that the team would even make the short list, he had been instructed by his principal to only put in the minimum amount of time necessary to allow the team to differentiate between the two possible bridge types. Therefore, he used assumptions for the required seismic loads for a similar bridge that he had helped design in the past. As it turned out, the seismic code had changed since that design was completed but he had not taken the time necessary to sort out the details of that change. He felt it was not necessary to "get it spot on" at this point in the competition; he only needed to get the team "in the ballpark."

The design-builder countered that while all those conditions were true, when the team did make the short list and was asked to submit a full technical and price proposal, the engineer did not revisit the conceptual design for the main span and told the design-builder's estimators to use the quantities that had been generated for the qualifications package. The engineer's response to this was that he did not want to expend more design effort than necessary to win the contract and that it was the design-builder's responsibility to cover scope creep with a pricing contingency. Also, since the design-builder had built many of these types of bridges, it should have known that quantities would grow during the design phase. After much wrangling of this nature, the two parties agreed to first apply the remaining contingency included in the contract to the overrun, and the engineer would pay for the remainder as a design error claim against its professional liability policy. When this claim was forwarded to the insurance company, the latter refused to pay, citing the same defense as detailed above.

The courts decided that the engineer was indeed liable for the damages caused by the Phase 1 design error and directed the insurance company to pay the entire amount of the damages. It cited case law that found that a design professional furnishes an implied warranty on any design information it conveys, regardless of the stage of design completion. The courts indicated that the design-builder indeed had the right to rely on the work of the engineer because the way the procurement was structured required it to submit a firm, fixed price for the bridge before the design was complete. Therefore, it was the engineer's responsibility to notify the design-builder as soon as he determined that that pre-award design was defective and to give the design-builder an opportunity to mitigate the damage.

Lessons Learned

The design process begins when the designer generates its first design product. In DB that happens before contract award, and contract award entails committing to a lump-sum amount prior to design completion. Therefore, design professionals on a DB teams must approach the design that they develop to allow the team to submit a responsive proposal just as seriously as when they are designing a facility that they know will indeed be constructed. The winning proposal becomes incorporated into the DB contract and, as such, the details of design included in it must be translated into the final constructed product. The risk for design errors and omissions is allocated to the design-builder. If it retains a design professional on a design subcontract, it must be able to rely on the competence of that person and the quality of his or her product.

Case 5-2: "Actions Speak Louder Than Words."

A public agency awarded a design-build-operate-own (DBOO) contract to a developer to furnish a large, multi-storey office building on a long-term lease arrangement in the desert southwest of the United States. The contract required the building to be designed and constructed in accordance with the agency's design standards and criteria. The lease included performance terms for various aspects of the building's operation. It allowed the agency to reduce or withhold its monthly lease payment for those months when the building did not meet the performance criteria. One of those criteria was the following requirement for the environmental control system:

> The ambient air temperature inside the building shall be regulated to maintain a temperature operating range of 72 °F to 75 °F with external temperatures ranging from 0 °F to 120 °F.

The developer chose to individually subcontract every design discipline necessary for the building, including the architecture and mechanical engineering. It did the same thing for all the construction trades, acting as its own general contractor/construction manager. Construction was completed ahead of schedule in December, and the agency occupied the facility in February. In April, the agency notified the developer that about one-quarter of the building was not meeting the room temperature criterion and, if the developer did not correct the problem, it would withhold 25% of its next monthly lease payment. The developer was unable to correct the problem, and by July the agency was withholding its entire monthly payment. It took two months to redesign and install a system that could maintain the required performance. The developer then filed suit against the mechanical engineer for the damages it experienced as a result of the mechanical system's failure to achieve the contract performance criteria.

Pursuant to preparing the defense for the lawsuit, the engineer's attorney determined the following facts about this case:

- The defendant was actually the second mechanical engineer to work on the design. The first had completed its design for the system and, in an agency review, it had been found to not comply with the agency's design standards. A dispute arose between the developer and the engineer over additional compensation because the engineer had not been provided the agency's standards, and the developer had fired the first engineer.
- The developer then directly hired the defendant to correct the deficient mechanical design and to inspect the mechanical construction, which it did. Its work consisted of converting a number of flexible ducts in the original design to rigid ducts to comply with an agency standard that runs of flexible ducts not be longer than 5 ft. It also reviewed the mechanical submittals from the trade subcontractors.
- During construction, the mechanical subcontractor found that the space the architect had allocated inside the building for one of four air handling units (AHUs) was inadequate; after coordination with the developer, the mechanical subcontractor relocated it outside on the building's roof.
- The reason why the rooms in the zone serviced by the relocated AHU could not maintain the required temperature was that the AHU had been sized assuming that it would be located inside the building's envelope. Correcting the issue required a new and much larger AHU to be installed on the roof and properly insulated.

The developer had used standard American Institute of Architects (AIA) contract forms for both the design subconsultants and the construction subcontractors. Each of the design contracts contained the phrase "The owner will furnish the necessary design information for the designer to complete its portion of the final design." In this case, the owner was the developer. No single design subconsultant had been designated as the DOR, and none of the design contracts required any of the designers to deconflict the work of all the other disciplines. In court, the defendant argued that because the developer was the only entity that had privity of contract with the other designers, it had constructively assumed the role of DOR and, as such, the error in the architectural design that created the conflict and resulted in the damages should have been found by the developer itself.

The developer's attorney countered that because the signatory on the contract was in fact a high school drop-out with no formal education in either architecture or mechanical engineering, it did not have the capacity to accomplish that task and, thus, the engineer had no right to rely on the contract clause regarding the owner's supplying the "necessary design information for the designer to complete its portion of the final design."

Outcome

The court found for the second mechanical engineer, stating that because the developer had failed to designate one of the licensed design professionals as the DOR, it had assumed that role was the only entity on the project that had the contractual capacity to coordinate all the designers. It pointed to the clause that stated that the owner would supply design information to each designer as the authority for making that judgment. Thus, by structuring the subcontracts for design and construction for the DB project in the manner in which they were executed, the developer had assumed liability for not only the adequacy of the design work, but also the performance of the constructed product.

Lessons Learned

If the owner wants to transfer design liability in a DB contract, it must guard its actions after award to ensure that it does not constructively take the liability back. In this case, the owner (i.e., the developer) tried to reduce the cost of design by not making the architect function as the DOR. In doing so, it usurped the design liability it thought it had transferred to the design professionals through their design contracts. While this is an extreme case, an owner can unintentionally accomplish the same unfortunate outcome by merely interfering with the design process and interjecting *directive* changes to the design in its design review comments. If the directive design changes are found to be the source of a design or construction error, the owner will have no recourse against the design-builder to make corrections.

If the developer in this case had not fired the first mechanical engineer and had required that person to make the necessary ductwork corrections, it might have been able to hold that engineer responsible. This is because the first mechanical engineer had furnished the architect with the space requirements for all the mechanical equipment and had personally participated in a "pre-final design review meeting" led by an agent of the developer. In that meeting, all the design professionals essentially reviewed and approved the final design as ready to release for agency review. The second engineer had been furnished the ductwork drawings only when it made the corrections, and in fact never looked at the full set of final construction documents.

The lesson here is that during DB project execution, it is important to both the owner and the design-builder to keep the design team intact. If a situation arises where this is impossible, the owner and the design-builder must take affirmative steps to see that design liability is transferred from the departing designer to its final replacement. Design liability is best managed by design professionals, and financial or technical issues should not be permitted to shift this liability to another entity in the DB project.

Case 5-3: "The Growing Design"

A public transportation agency had planned to deliver a project using traditional DBB, but during the design it encountered a number of thorny environmental and right-of-way issues. It therefore suspended the design contract at approximately 50% design completion. During the suspension, the consultant's design project manager left the firm and went to work for another engineering firm. Once the issues that halted the project had been sorted out, the owner then decided to award the project as a DB contract to recover some of the lost time, and specified a very aggressive delivery schedule. It believed that the schedule was possible because of the advanced state of the "preliminary" design, which were called "bridging documents" in the RFP. The contract required the design-builder to validate the technical acceptability of the bridging documents and assume liability for integrating those portions that were found to be acceptable into its own final design.

One of the competing design-build teams had partnered with the engineering firm that had hired the preliminary design consultant's former design project manager. The owner was asked if this constituted a conflict of interest that would prevent the design-builder from assigning this individual to this project. The owner answered that it was not; in fact, during evaluation of qualifications, this individual's previous knowledge of the project was rated as a distinct advantage for that DB team. It went on to win the contract.

The design-builder was a general contracting company that hired the DOR on a design subcontract. The teaming agreement that was executed during proposal preparation required the DOR's firm to provide the following services:

> Given the [design-builder's] reliance on the input from the [DOR's] design team to help develop the proposal, presentation and bid, the [DOR] will perform supporting services including designs and analyses for project pursuit in excess of those normally needed for "pure design" services proposal.

Part of those services included generating quantities of work and furnishing the design-builder's estimating staff with appropriate parametric design formulae for those features of work it deemed to be routine and therefore did not require additional design effort. An example of this type of formula was one where the engineer specified that the estimator allow X pounds of reinforced steel for every square yard of standard DOT reinforced concrete pavement.

After award, the design work was advanced very quickly and the major features of work were released for construction ahead of schedule. During a cost engineering review of the feature of work that was added to mitigate the project's environmental impact, the estimator found that the quantities of excavation associated with these features had tripled from the quantities used in the price proposal. Additionally, stone riprap quantities had also significantly increased. The riprap had been estimated using a formula of X tons per square yard of excavated surface area. A meeting was held with the design project manager to investigate the cause of these quantity increases and seek potential solutions. The design project manager indicated that, due to his previous familiarity with the bridging documents and the advanced stage of the design in the RFP, he had not needed to check the quantities that were already shown on the preliminary design documents. He had taken them and added a 10% design allowance for possible growth during detailed design in the quantities he furnished for the preparation of the "bid." Analysis of the revised environmental mitigation plan after award found that certain features were made larger to comply with the permit, and that this was engendered by acquisition of a previously unavailable right-of-way. He had not known that the agency had been able to get the land and hence assumed that the environmental mitigation features were constrained to the pre-proposal project limits shown in the bridging documents. The RFP contained a new drawing that showed the current project limits, and its impact had not been evaluated during development of quantities for the DB price proposal.

The design-builder took the position that this was the DOR's mistake and therefore it should be liable for the increased costs attributed to the growth in excavation and riprap quantities. The DOR disagreed, citing the following defense:

- The design-builder knew that the team had only a limited time to prepare its proposal, and they had agreed that the design team would invest only the minimum level of effort required to submit the proposal.
- The design project manager had included a design allowance in the as-bid quantities; the fact that it was exceeded was a normal risk in doing DB contracting, which the design-builder should have accounted for because it had extensive experience with DB delivery projects of this nature.
- The design-builder should have included a sufficient contingency in its price proposal to cover quantity growth during design.
- Due to the aggressive schedule, there was no time allotted for stopping the design to check the current project cost, and the design-builder should have taken that into account when it set the project's profit margin, which it believed to be too low.

Outcome

This dispute was settled in binding arbitration. It involved only the DB team, not the owner. The panel found that because the design-builder was not a licensed design professional, it did not have the capacity to understand the magnitude of potential quantity growth during design. The panel agreed that it was the engineer's responsibility to add a design allowance for quantity growth for each item that it supplied in either the design or the quantities themselves to arrive at the as-bid quantities. Because the design project manager failed to take into account the fundamental changes to the project's scope of work between the time he had previously worked on it and when the proposal was actually submitted, the DOR was guilty of an act of omission. Additionally, the design project manager failed to keep the estimating staff up-to-date by submitting a design product for cost engineering review before sending it to the owner for review and release for construction. The argument that the majority of the as-bid quantities on the rest of the project were accurate did nothing to ameliorate the fact that a significant change had been made to the bridging documents, and the design team failed to identify it during the development of as-bid quantities. Finally, the terms of the teaming agreement indicated that the design-builder would rely on the input developed by the DOR in the preparation of its bid. Therefore, the engineer who had the capacity to accurately estimate the potential quantity growth should have communicated the change to the bridging document quantities to the design-builder with its as-bid quantities.

Lessons Learned

It is important for designers in DB projects to fully understand the impact of the fact that they must design to both a budget and a schedule. Designers typically do not have the intimate familiarity with construction market conditions that constructors do. Thus, knowing that the project has a fixed monetary amount of contingency is of little use to a design engineer or architect. What is cogent is the knowledge of the total amount of as-bid quantities of work for each feature that is being designed and using that knowledge as a budget to guide the design effort. Designers should be instructed to review the quantities on a regular basis and, if the as-bid amount is going to be exceeded, to bring that to the immediate attention of the DB team. Then the team can review the reasons for the quantity bust and either change the design to bring it back on budget, or cross-level the contingency from items that are under-running their quantity budget. As a last resort, they can try to mitigate its impact by making changes to features of work that have not yet been designed.

Case 5-4: "All the Right Moves"

The RFP for a new bridge that was meant to replace an old bridge stated "The alignment of the new bridge shall not change from the existing alignment." A 60-year-old geotechnical report for the project site was provided to "furnish a representative condition on which to base preliminary foundation designs." The RFP also required the design-builder to execute soil borings and provide a final geotechnical report "on which he will base his final design." The design-builder's price proposal indicated that the use of piles was contemplated for the foundations of both abutments. The contract also required the design-builder to design the north end first because it was the more complex design and had the least available real estate on which to place the northern footprint.

During the owner's review of the southern foundation design, the owner received a working drawing indicating a shift in the alignment for the south end of the bridge to the edge of the existing right-of-way within the designated project limits, a distance of about 35 ft. Upon questioning the geotechnical engineer, the owner was shown the required final geotechnical report that showed that a shelf of bedrock was located on that side of the project site, which apparently had not been found on the original survey for the old bridge. This condition permitted the design-builder to install one abutment on a shallow foundation system.

The design-builder maintained that the RFP required him to redo the geotechnical survey and, because he furnished a more extensive geotechnical survey than what was probably contemplated by the authors of the RFP, he was entitled to reap the benefit of the cost savings allowed by this discovery. Furthermore, he chose to interpret the clause regarding alignment to mean that the bridge must be built on the available property described in the site plan. Additionally, in order to effect the increased capacity requirement for the new bridge, he was forced by project geometry to slightly alter (by about 5 ft) the alignment on the north side of the bridge to accommodate new access ramps within the existing project limits. This change had been noted by the owner during the review of the north end and was accepted as the only possible solution to keep from delaying the project by having to procure the necessary new right-of-way. Therefore, if it was acceptable to move one end of the bridge within the project limits to make it work, it should follow that the shift from the existing alignment on the other end would be acceptable as well. Finally, the design-builder added that moving the south end of the new structure allowed him to maintain a full and unimpeded flow of traffic during construction for an additional six weeks. The schedule showed that closure would have occurred at the beginning of the tourist season and this bridge was on a main route to a major tourist attraction in the area. Thus, the owner was receiving the value of decreased user costs of construction during the life of the project.

After much internal legal and technical discussion, the owner recognized the advantages of the new alignment. However, the owner felt that a credit under the contract's value engineering (VE) clause should be offered by the design-builder due to the cost savings from the original price proposal, because the owner was furnishing relief from the "shall not change from the existing alignment" RFP performance requirement. The design-builder refused to entertain the idea of sharing the savings, stating that it had exceeded the RFP performance criteria for the geotechnical survey and, as a result, should receive the windfall associated with conducting a much more thorough geotechnical study. The design-builder also argued that the shift in the north end of the bridge to accommodate the access ramps actually exceeded the price for that feature that was contained in the price proposal. By this time, the project was beginning to suffer a schedule delay from the inability to resolve this particular issue. The dispute was referred to a mediator.

Outcome

After a short meeting, the parties agreed that in spite of the owner's RFP language and the design-builder's price proposal, the requirement to conduct a detailed geotechnical investigation and base the design on the final geotechnical report left open the final design and allowed the DB contractor to accrue the benefit of a less expensive foundation *if* the facts supported that design decision. The owner conceded that the previous approval of a "change" from the original alignment probably constituted a waiver of that RFP requirement. The design-builder agreed to not make a claim for the time delay caused by this dispute and to accelerate the work to make up the schedule.

Lessons Learned

An understanding of the disposition of "minor" changes to the contractual scope of work as a result of design reviews is what must be learned from this case. The owner's acceptance of the minor shift at the north end rendered inoperable the disputed RFP clause regarding the bridge's final alignment in this case. Thus, the VE clause could not be invoked because there was nothing left in the contract to change and thereby render a tangible savings to justify the change in design. If, however, the owner had recognized the possible impact of that very logical and sound decision to allow a shift in the alignment on the north end and had executed a no-cost change order to reword the alignment clause to reflect the north end shift, then it would have contractually "pegged" the south end to the original alignment. Then the owner would have been entitled to the credit under the VE clause. It is important for the owner's design review staff to be cognizant of the impact of allowing minor variations and deviations from the contract scope of work and ensure that appropriate contract actions are taken to codify them in the contract as they occur.

Case 5-5: "Keep Looking until You Find the *Right* Answer."

A DB RFP stated that all buildings over a certain height at an installation "shall be founded on caissons." A 6-year-old geotechnical report for an area two miles away from the project site was provided to "furnish a representative condition on which to base preliminary foundation designs." The RFP also required the design-builder to execute soil borings and provide a geotechnical report "on which he will base his final design." The design-builder's price proposal indicated that the use of caissons was contemplated for the foundation.

When the preliminary geotechnical report was submitted to the owner, it contained three reports from three different subconsultants:

- Report 1 stated: "Building A must be founded on caissons."
- Report 2 stated: "This report was prepared to determine if Building A really must be founded on caissons. Report 1 used soil tests that are normally used on loam soil, and the soil in question is clay. Nevertheless, if the results were used to determine the requirement for a deep foundation, they would indicate caissons, but as the soil type is different, Report 1 may be incorrect in its application of test data to foundation design."
- Report 3 (prepared by the DOR's structural designer and in-house geotechnical engineer) stated: "Having reviewed the information

contained in Reports 1 and 2, we disagree with their conclusions. We agree with Report 2's assertion that the tests results were not properly evaluated with respect to clay soils. After additional testing and an analytical study of all the results, we find that Building A should be founded on spread footings."

Exercising the prerogative to "proceed at his own risk," the design-builder began preparation to put Building A on spread footings before the final geotechnical report was submitted. The owner's experience in this area indicated that certain areas of the base required deep foundations and others did not. This project was literally on the geographic boundary between the two areas. Additionally, the owner had awarded this project at a price that virtually eliminated any contingency funds for the owner to add scope to the project. The issues in this case that must be analyzed are:

- Did the owner have the authority to require the contractor to suspend foundation work until the final geotechnical report was submitted?
- If the owner believed that a shallow foundation brought an unacceptable risk of performance with respect to settling, could the DB contractor be required to build caissons without a compensable change order?

Outcome

In spite of the owner's RFP language and the design-builder's price proposal, the requirement to conduct a detailed geotechnical investigation and base the design on the final geotechnical report left open the final design and allowed the design-builder to accrue the benefit of a less expensive foundation *if* the facts supported that design decision. The owner had to decide whether the risk of settlement and its attendant headaches associated with trying to recover damages was worth less than negotiating an adjustment to require caissons. In this case, the owner commissioned an outside geotechnical expert to prepare a report based on the contractor's soil data. The report showed that both methods appeared to be adequate but that spread footings had a low factor of safety (1.25). As a result, the outside expert recommended that caissons be used. A settlement was reached where caissons were designed and installed in exchange for a reduction in scope in a later feature of work. Thus, both budget and technical requirements were maintained. However, a time extension for the preparatory rework was granted.

Lessons Learned

The way this preliminary report was prepared certainly brought the quality of the design-builder's design process into question. It appeared on the surface to be an attempt to find an engineer who would authorize the less expensive foundation and allow the design-builder to increase its profit margin. This type of transaction assassinates the required level of professional trust that is necessary to successfully execute a DB project. The fact that the design-builder included the first two reports probably indicated that it was in fact being totally honest with the owner. However, the lesson to be taken here is that the design-builder should have reconciled the differences between the three opinions and determined the best course of action. That course should have been in the geotechnical deliverable that it submitted to the owner. In fact, it is clear that the conservative decision would have been to carry forward with the caissons and not attempt to save any costs.

SIX

Construction Quality Management in the Design-Build Project

In design-bid-build (DBB) projects most owners are accustomed to describing in great detail the means and methods used to carry out the construction in the standard specifications for construction. These specifications have been used in traditional project delivery and have proven over time to be successful in yielding a quality product. In DB, however, owners have the opportunity to allow design-builders to use specific construction means and methods to differentiate themselves from their competitors and to provide efficiencies that may not have been contemplated by the project's owner. Yet with this opportunity also comes the risk that the means and methods used by the design-builder may not achieve the same quality as those prescribed in DBB contracts. Thus, one of the major quality management (QM) decisions that must be made in a DB project is to articulate the amount of flexibility the design-builder will have over construction means and methods.

If one were to ask a room full of DB practitioners to write down their definition of the phrase "construction quality," one would get a different definition from each respondent. Table 6-1 presents a typical breakout of the quality functions and the entities responsible. This is often the basis of discussions that result in the total quality program for the organization. The trend, however, would be that responses from practitioners from the same discipline (e.g., design, construction) would generally group by perspective. Thus, it is important for both the owner and the design-builder to be using the same quality definitions during construction.

Here are examples of different definitions that spring from the definer's perspective:

- Design-oriented: "Quality consists of the extent to which a specimen [a product-brand-model-seller combination] possesses the service characteristics you desire." (Maynes 1976, p. 542)
- Price-oriented: "Quality refers to the amounts of the unpriced attributes contained in each unit of the priced attribute." (Leifler 1982, p. 956)
- Customer-oriented: "Quality consists of the capacity to satisfy wants." (Edwards 1968, p. 37)

Table 6-1 Typical Construction Quality Management Tasks in Design-Build Projects[a]

Entity	Quality Control (QC)	Quality Assurance (QA)	Project Quality Assurance (PQA)
Owner	None, unless owner furnishes construction with own forces to project.	Checking of pay quantities Audit QC testing Review and approve construction submittals that were reserved for owner approval (e.g., architectural finishes, color selection)	Approval of post-award QM/QA plans Audit design-builder's QA plans and activities Independent verification testing Acceptance testing Oversight Approval of construction progress payments Final acceptance inspections
Designer-of-Record	Technical review of shop drawings Technical review of material submittals Prepare and check as-builts	QA inspection QA testing Verification/acceptance testing Approval of post-award QC plan Review all O&M manuals for compliance	
Constructor	Construction QC planning Checking of pay quantities Routine construction inspection QC testing Establishment of horizontal and vertical controls on site	Review construction submittals from subcontractors Monitor compliance with approved submittals	
Trade Subcontractors	Prepare and submit construction submittals Build to approved designs		

[a] The lists in this table are not meant to be all-inclusive.
O&M, operation and maintenance.

- Contract-oriented: "Quality is any aspect of a product, including the services included in the contract of sales, which influences the demand curve." (Dortman and Steiner 1954)
- Market-oriented: "In the final analysis of the marketplace, the quality of a product depends on how well it fits patterns of consumer preferences." (Keuhn and Day 1962)
- Performance-oriented: "Quality is fitness for use." (Juran 1988, p. 2)

In DBB, many public owners and private design firms have a standardized set of specifications they are comfortable using on a repetitive basis. In transportation, state DOTs usually make their standard set of specifications for bridges and roads a contract requirement. Thus, as discussed in Chapter 4, it is difficult not to routinely prescribe that the design-builder use the traditional set of technical documentation during the design phase. When shifting the project focus to construction-phase quality, it must be remembered that there is often a different set of professionals who will administer the construction QM program at the project site than those who were involved in the design QM process. Research has found that most DB project owners have well developed construction QA/QC programs for their DB projects (Gransberg and Molenaar 2004). However, many of them depend on prescriptive measures of quality and are not designed to oversee construction quality when it is specified in performance terms. Therefore, it is exceedingly important that owners review their standard DBB construction QM programs and ensure that they are adjusted to account for the characteristics of DB project delivery. The DB project quality assurance (PQA) model was introduced and discussed in detail in Chapter 4. Figure 6-1 shows how it relates to the construction QM process. Additionally, Table 4-1 in Chapter 4 lists the construction QM tasks broken out into QA, QC, and PQA. The use of these two models will be continued in this chapter.

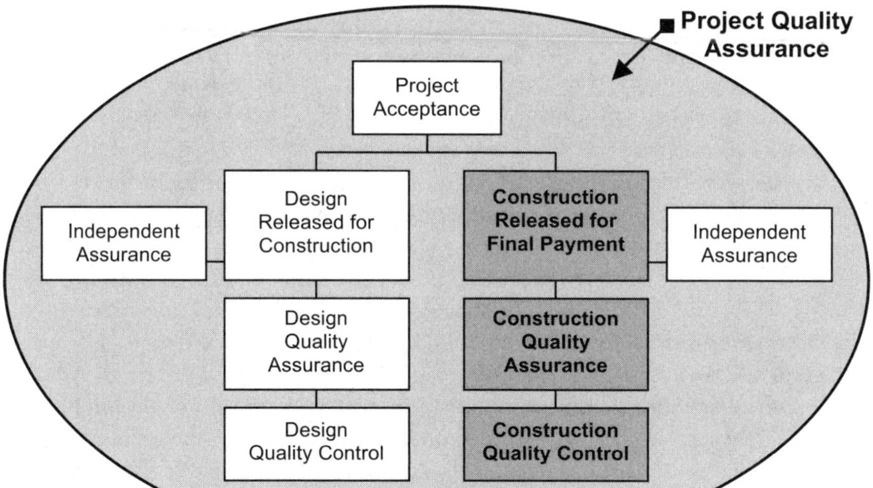

Figure 6-1 DB project quality assurance model (construction phase).
Source: Gransberg et al. (2008), Figure 3, with permission from the Transportation Research Board of the National Academies.

Construction Quality Defined

To review the definition provided in Chapter 4, "quality" is defined as follows:

> (1) The degree of excellence of a product or service; (2) the degree to which a product or service satisfies the needs of a specific customer; or (3) the degree to which a product or service conforms with a given requirement. (TRB 2005)

Given the three possible definitions as the basis for defining design quality, it would seem that the last one, conformance to a given requirement, best fits construction quality in the DB context. In DB construction QM, the "requirement" with which the constructed product must conform is the approved design documents. Thus, in a fast-track DB project where construction begins before design is complete, the challenge is to identify the requirements in a timely manner that allows the construction to proceed without interruption.

The American Society for Quality (ASQ) furnishes some additional definitions that are useful during the construction phase of a DB project. These describe different kinds of quality:

- Relative Quality: Loose comparison of product features and characteristics.
- Product-Based: Quality is a precise and measurable variable and differences in quality reflect differences in quantity of some product attribute.
- User-Based: Fitness for intended use; need to know frequency of use, cost, performance, reliability, and serviceability requirements.
- Manufacturing-Based: Conformance to specifications.
- Value-Based: Defined in terms of cost and price; performance at an acceptable price or conformance at an acceptable cost. (ASQ 2008)

It can be seen that there will be components or features of each of these five types of quality, regardless of DB project type. For instance, components that must conform to manufacturing-based quality standards will be found in the approved specifications. A performance specification that contains a product "or equal" clause furnishes a relative quality benchmark. An RFP performance requirement to deliver interior finish treatments on a building that cost no more than a given percentage of the total cost is an example of a value-based quality criterion. A concrete material specification that dictates the concrete's minimum compressive strength is a product-based quality criterion. It thus follows that the owner and the design-builder need to communicate with each other about their expectations and their interpretation of quality in each project feature.

Types of Construction Quality

ASQ defines relative quality as "comparison of product features and characteristics, which is determined by the attitude of those who are involved in the work. The design communicated by the specification establishes the relative quality of the materials and workmanship" (ASQ 2008). This is sometimes called "transcendent quality," which indicates that there will be difficulty in expressing a clear specification of quality. However, the user will "know" when it is good enough. In *Zen and the Art of Motorcycle Maintenance*, R. M. Persig describes it this way: "Quality is neither mind nor matter, but a third entity independent of the two . . . even through Quality cannot be defined, you know what it is" (Persig 1974, pp. 185–213). The historian Barbara Tuchman used the term "excellence" to qualify the definition of quality: ". . . a condition of excellence implying fine quality as distinct from poor quality. . . . Quality is achieving or reaching for the highest standard as against being satisfied with the sloppy or fraudulent" (Tuchman 1980).

One can see the potential for disputes when expectations are not clearly defined but, rather, rely on the design-builder to furnish the "highest standard" instead of an explicit minimum standard. This definition is often used to express intangible requirements for things like architectural approach or a desire to minimize environmental impact. No two architects will have exactly the same opinion regarding architectural aesthetics and, no matter how "green" a design-builder attempts to be, a passionate environmentalist will always be able to find a way to further minimize environmental impact. Thus, the use of relative quality to set a standard should be avoided when possible.

Product-based quality is very tangible. Lawrence Abbott, an economics professor, said, "Differences in quality amount to differences in the quality of some desired ingredient or attribute" (Abbott 1955, pp. 126–127). Hence, this definition entails the ability to measure various attributes of the product, and this ability reduces the potential for dispute in a DB project. The issue here becomes its constraint of potential innovation and a focus on achieving the minimum level of quality, whereas the definition of user-based quality depends on a detailed understanding of the user's postconstruction requirements. Put another way, "[User-based] quality is the degree to which a specific product satisfies the wants of a specific consumer" (Gilmore 1974, p. 16). This quality definition requires the owner to have developed a very clear picture of user requirements prior to advertising the DB contract. If this has not happened, this definition cannot be used in a DB project.

The definition of manufacturing-based quality is also the definition of a prescriptive specification. "[Manufacturing-based] quality is the degree to which a specific product conforms to a design or specification" (Gilmore 1974, p. 16). Therefore, this definition should only apply to features of the DB project where there is only one acceptable technical solution, and the owner must reckon with the assumption of potential design liability for the performance of those features.

Finally, the value-based quality definition creates an opportunity to allow the maximum amount of creativity within the constraints of the DB project's budget. "[Value-based] quality is the degree of excellence at an acceptable price and the control of variability at an acceptable cost" (Broh 1982, p. 3).

Given the various ways to define quality, it is important to ensure that all construction QM plans include mechanisms for distinguishing acceptable quality for the given feature of work. This happens through the owner's review and approval of the design-builder's QM plans.

Owner's Role in Defining Construction Quality Requirements

Owners must remember that they give up control over the details of design by selecting DB project delivery, and thus are depending purely on the qualifications process to guarantee project quality. Because that control is transferred to the design-builder, the constructor will build what the designer has designed. Additionally, because the constructor and the designer are now on the same team, there is no need to produce biddable construction documents. This leads to some difficulty for the owner's QM personnel because there may not be an explicit set of construction details that are part of the approved/released for construction plans and specifications against which to measure the construction quality. Therefore, it is important that the winning DB team's approach to construction QM be evaluated prior to contract award. Because construction quality requirements flow out of the design documents of a DB project, the owner must clearly articulate its contractual definition for construction quality during the design phase.

DB RFPs usually provide a global definition of quality that usually does not directly address specific elements of construction quality. This requires the owner to continually provide clarifications to the contractual definition that are consistent with the definition that was interpreted and bid by the design-builder prior to award. A clear definition of construction quality is imperative for every project. If it was not clearly defined prior to contract award, both the design-builder and the owner must develop a clear and measurable definition that is agreeable to both parties prior to starting detailed design.

The owner's role in defining construction quality requirements consists of the same three major areas as in design QM:

- Establish and communicate construction quality objectives.
- Ensure compliance with both the contract and the implicit statutory requirements applicable to the project.
- Define the roles of the various contributors to the construction QM program.

The first two bullets flow directly from the owner's review and approval of the design-builder's design documents and their subsequent release for construction. The last bullet is realized by the approved QM plans for construction.

One simple method for ensuring that QM roles and responsibilities are fully understood is the use of a linear responsibility chart. Figure 6-2 and Table 6-2 provide examples of how this could be applied to construction quality management. One can see that all the players are listed across the top of the matrix and the various QM activities are listed down the left side, with each entity's responsibility indicated.

Responsibility Codes	Organization								Design-Build Project
A. Intiation/Action B. Review C. Approval D. Information Only N. Not involved									
Construction Quality Management Activity	Owner	Preliminary Design Consultant	Design-Builder QM	Designer-of-Record QA	Constructor QC	Trade Subcontractors	Independent Material Test Lab	Quality Auditor	Remarks
QM/QA/QC plans	C	N	B	B	A	D	D	D	
Audit design-builder's QA plans and activities	D	N	A	A	A	A	A	C	
Independent verification testing	D	N	D	D	D	D	A	D	
Verification testing	D	C	B	B	A	A	D	D	
Acceptance testing	C	A	B	B	B	D	A	D	
Review/Approve owner-reserved submittals	C	B	B	B	B	A	N	D	
Review/Approve construction submittals	D	D	B	C	A	A	N	D	
Technical review of shop drawings;	D	D	B	C	A	A	N	D	
Technical review of material submittals;	D	D	B	C	A	A	A	D	
Prepare and check as-builts	C	B	B	A	A	A	N	D	
Routine construction inspection	D	D	B	B	C	A	N	D	
QC testing	D	N	B	B	C	A	B	D	
QA testing	D	D	C	A	D	D	B	D	
QA inspection	D	D	C	A	D	D	B	D	
O&M manuals compliance	D	D	B	C	B	A	N	D	
Punchlist	C	B	B	B	A	A	N	D	
Final acceptance inspections	C	B	B	B	A	A	N	D	

Figure 6-2 DB construction quality linear responsibility chart.
O&M, operation and maintenance.
Source: Adapted from Clark (2007).

Construction QM in DB projects is strikingly similar to its counterpart in DBB, but there are two major differences. First, the proportion of performance criteria and specifications will normally be greater in DB because an owner who chooses DB usually also wants to leave the door open for innovation by the design-builder and its team. Secondly, in DB the probability that construction will be undertaken using partial design documents is high. The major reason why owners choose DB project delivery is to compress the project delivery period (Songer and Molenaar 1996). The most direct way to accomplish this objective is to permit construction to begin as soon as possible and not wait for 100% design completion. Therefore, the remainder of this section will be devoted to discussing the mechanics of performance quality objectives and undertaking construction QM in an environment where the design is not complete.

Performance Quality Objectives

Our previous book developed the following definition for a performance criterion:

> A rule by which the effectiveness of an operation or function is judged and its value measured. (Gransberg et al. 2006)

Table 6-2 Example of Construction Quality Assurance Responsibility Distribution

Item	Owner	Designer-of-Record	Builder	Trade Subcontractors
QA/QC System	Defines overall system	QA/QC for design QA for construction	QA on Subs; QC for construction	QC for work; QA on suppliers
Administrative Submittals	Review and approve	Review for RFP compliance	Submit and conduct internal checks	Submit and comply
Shop Drawings	Information only	Review, deconflict, and approve	Review, build, and monitor	Submit and build
Equipment and Material Submittals	Information only	Review, deconflict, and approve	Review, build, and monitor installation	Submit, furnish and install
Architectural Submittals	Review and approve	Review for RFP compliance	Review and build	Submit and build
Mock-ups	Review for RFP compliance	Review and approve	Review and check	Furnish and install
O&M Submittals	Review for RFP compliance	Review and approve	Submit and verify	Submit and verify
Close-out Documents	Review and approve	Review for compliance if necessary	Submit and verify	Submit and verify

RFP, Request for Proposal; O&M, operation and maintenance.

So, keeping in mind that the operative words in the definition are "effectiveness of operation or function," "judged," and "measure of value," the owner can now establish a set of quality objectives that correspond with the performance criteria contained in the DB contract and articulate the "rule." The development of appropriate QM plans for construction should be a collaborative process. This process can be highly structured and is commonly called "quality function deployment" (QFD). Pheng and Yeap describe it as a "quality improvement technique that deals with quality problems from the outset of the product design and development stage and assures that customers' requirements are accurately translated into appropriate technical requirements and actions" (Pheng and Yeap 2001). Figure 6-3 illustrates how QFD is applied in a DB project. One can see that it starts with a merging of the owner's and design-builder's standard QM plans to produce a QM plan that is tailored for the project at hand. The flow chart does not differentiate between design and construction where QFD is concerned.

The quality performance requirements that are applicable to design are also applicable to construction. Thus, the standards established by the design process must necessarily be implemented in construction to furnish the desired performance requirement in the completed project. Figure 6-3 also shows that this is a cyclic process for both the owner and the design-builder. To attain continuous

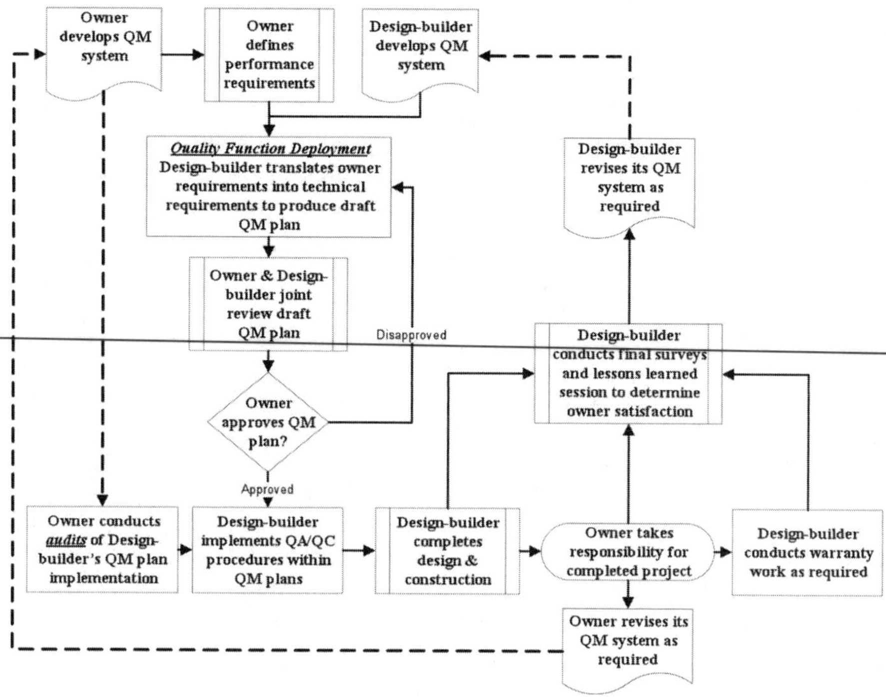

Figure 6-3 Quality function deployment in the DB project.

improvement, both must capture lessons learned and feed those back into their standard QM plans to avoid making the same mistakes on the next project that were made on this one. Thus, success is defined by the owner clearly communicating its quality performance objectives and the design-builder converting them into implementable QM plans that flow through design into construction and finish at the end of the postconstruction warranty phase.

QFD, in a nutshell, involves the design-builder taking the owner's technical requirements and performance objectives and developing technical design requirements and associated construction QM plans. The design-builder then determines QM techniques that will provide a comfort factor to the designers, constructors, the owner, and other stakeholders. It then determines how its internal QM system supports the owner's QM system, and creates interface points where the two can be merged. Typical interface points are:

- QA plans must be synchronized with QC plans.
- Plans and specifications conform to RFP performance criteria.
- Shop drawings conform to plans and specifications and are integrated into the overall design.
- Subcontractor submittals conform to plans and specifications.

- Construction placement and installed equipment conform to specifications and plan/shop drawing dimensions and materials.

Finally, the design-builder integrates quality plans and procedures of all its team members into a comprehensive QA/QC plan which can be audited as required by the owner's QM personnel to verify that it is being implemented as approved.

Quality Management with Incomplete Design Documents

The fundamental challenge for making construction QM plans workable in the DB environment is forming them with the idea that QM personnel must be able to implement them when construction documents are either incomplete or not finally reviewed and approved. This is necessary to allow the owner to achieve its goal of delivering the project in the shortest possible or practical time. To do this successfully requires a good deal of planning and analysis by the owner's QM staff.

The first step is for the owner to break down the project into its component features of work, as described in Chapter 2. Each feature should be examined regarding its individual quality risk. Features with a high risk of unacceptable construction quality must be identified as features that cannot be constructed until the design effort for those features is complete. For those features where the quality risk is tolerable, the design-builder can attempt to accelerate the schedule. Obviously, this will affect the design schedule, with the high-risk features receiving high priority for early design effort to mitigate potential delays due to incomplete designs.

The second step is to develop explicit QA/QC plans that contain a look-back feature to return to and inspect features of work built from incomplete construction documents to ensure that they indeed comply with the final approved design. The final step is to include a mechanism to adjudicate deficiencies identified during the look-back inspections in a manner that is fair to both the owner and the design-builder. Some owners have established a specific contingency fund for these purposes that can be tapped if the pace of construction outstrips the pace of design, giving the owner a project as quickly as possible but with some quality issues that must be corrected prior to final acceptance.

The Economics of Quality in the Construction Phase

The economics of quality in the construction phase are characterized by determining how much QM effort on the part of the owner and design-builder is sufficient to deliver a quality project. More inspection does not necessarily equal more construction quality. There is an optimum point at which the money spent on QM activities breaks even with the value added by those activities. It is unreasonable to think that anything as complex as a DB civil engineering or building project could be constructed with no defects. It is reasonable, however, to expect that defects that are identified will be corrected before the project is finally accepted. Thus, striking this balance becomes the formula for defining the construction quality economics.

Design-Build Construction Quality Issues

It is not the intent of this section to cover each and every aspect of construction quality on all types of projects. We approach the topic globally, addressing the major general issues that we have encountered in both the literature and our personal DB project experience. These can generally be listed as:

- Allocating QA responsibilities during construction.
- Resolving professional differences of opinion during quality audits and inspections.
- Scope creep during construction.
- Ability to change previously approved construction submittals.

Allocating Quality Assurance Responsibilities during Construction

It is the owner's prerogative to establish the QM system with which it feels most comfortable. Thus, it is the owner's duty to allocate construction QA responsibilities. Generally, there are two options that the owner's QM personnel can take. The first is to require the design-builder to develop the project-specific QM plans during QFD in accordance with the owner's standing QM plans and policy. The advantage of this option is that the owner's PQA crew will be familiar with the procedures and be better able to hit the ground running and facilitate project progress. The downside is that it forces the design-builder's personnel out of their comfort zone and requires them to dedicate time and resources early in the project to training the construction QM personnel.

The other option is to permit the design-builder to adapt its standing QM program to the owner's needs. This requires that the owner's QM personnel review the proposed plan in depth and ensure that an understanding of the standards, procedures, and processes is reached. This also entails ensuring that there are no false expectations regarding the responsibilities that the owner chooses to reserve for itself. Regardless of which option is selected, the important factor is that every stakeholder in the construction QM process both acknowledges and understands its roles and responsibilities before construction commences.

Resolving Professional Differences of Opinion during Quality Audits and Inspections

The most disconcerting and often-encountered issue regarding construction QM is a difference in professional judgment that occurs when one competent construction professional inspects the work of another. These issues most often occur when a feature of work that has been detailed using a performance specification or criterion is completed. This typically arises when the design contains an "or equal" specification (i.e., relative quality) and a trade subcontractor delivers a submittal that does not offer to provide the named product against which all other alternatives will be compared. The owner generally suspects that the substitute must be less costly and hence of lower quality. However, if it is indeed equal to the named product in performance terms, the design-builder or its subcontractors are entitled to incorporate it into the project. To preserve the element of trust that is so necessary to a successful DB project, the owner's PQA staff should withhold judgment and allow the design-builder's staff to present their case to prove that the relative quality of the substitute is satisfactory and conforms to the specification. It must be remembered that relative quality is a performance-based definition, not a prescription for exact duplication. In other words, if the substitute will perform equally to the named product, then it is acceptable even if it is not a precise copy.

The simple solution to resolving these kinds of issues is to take the time during design to select exactly what should be used in construction and not pass the buck to the trade subcontractors to make these decisions about extensions to the approved design. Because the construction contract is awarded simultaneously with the design contract, there is no need to attempt to stimulate competition. The competition ended when the DB contract was awarded, so utilizing an "or equal" performance specification is merely a relic from the DBB procurement culture that needs to be excised from the DB design and construction quality process.

Scope Creep

The risk of scope creep does not end when the final design is released for construction. Scope creep in construction happens during submittal review when the owner demands that work be incorporated into the final construction that was not contemplated during the preparation of the project's price proposal. Thus, it is often the result of a failed design QM process. If the design-builder's design QA staff becomes too accommodating (too responsive to the owner's PQA staff's requests and preferences), the released construction documents will portray a scope of work that exceeds the scope contemplated in the original price proposal. If this occurs, it is incumbent upon both the owner and the design-builder to admit that there is a problem and work together to recover the budget through thoughtful value engineering (VE).

Scope creep can also appear during routine QA/QC inspections and quality audits. In this case, it is generally caused by a conversational request to "slightly" alter the approved design to enhance it in some manner. An old saying is "Nickels and dimes add up to dollars," and if this type of behavior is allowed to go unchecked, the design-builder will find its profit margin eroded for no apparent reason. This type of scope creep is best controlled by a two-part system. First, at the pre-work conference, both the owner's and the design-builder's personnel are told exactly who can and who cannot make field changes to the design. A system for requesting field changes should be established and integrated into the overall project QM plan as appropriate in the QC, QA, and PQA levels. The second part of the system is a rigorous project controls system that tracks actual costs against the estimated costs and is responsive enough to raise red flags when costs begin to break a reasonable variance. The system should give warning in time to investigate the source of the cost performance problem and take corrective action to recover the budget.

The design-builder will also need to educate those who caused the problem to ensure it does not happen again.

Variances from Approved Submittals

Once a construction submittal has passed through the approved QM chain and is approved for incorporation into the constructed product, there should be no variations except in unusual or unavoidable circumstances. The cost and time it takes to properly review and approve a submittal is significant, and the design-builder should endeavor to develop a system via its QFD role that "gets it right the first time." However, no system—no matter how well planned and devised—will ever be perfect, and the QM plan should contain a provision and a sequence of events for requesting a change to an approved submittal. This usually involves the actual unavailability of an approved product or service; this could happen for a number of reasons, the foremost of which is market pressure.

For example, in the summer of 2008 the major supplier of asphalt products in Oklahoma went bankrupt, causing a short-term shortage of its product, which in turn caused the Oklahoma Department of Transportation to drastically alter its 2008 construction program (Tulsa World 2008). Many of the projects awarded in the July letting were put on hold until an alternate supplier of construction material could be found. Coincidently, this occurred simultaneously with the year's spike in oil prices, which made concrete pavement less expensive than asphalt pavement for a short period of time. Although this occurrence was not related to any DB projects (Oklahoma does not use DB contracting in its transportation program), it is a great example of a reason to permit a variance from a previously approved submittal. If one of the projects affected by this crisis had been a DB asphalt paving job, it would have furnished justification to change the design to concrete pavement as the only sure means to complete the project.

Another common reason for a variance is related to the rapid advance of technology around the world. Medical projects are particularly susceptible to these types of submittal change issues. Most healthcare projects involve the installation of state-of-the-art medical equipment, which can change between the time a contract is awarded and the construction of the medical facility is completed. One extreme instance occurred on a DB hospital project. A particular piece of medical imaging equipment was designed to be powered from a drop through the ceiling, and electrical shop drawings were developed to accommodate this requirement. After the submittals were approved but before the equipment was purchased, the technology changed and the resultant piece of equipment that was purchased was designed to be powered from the floor. This necessitated a substantial change in the electrical power distribution plans. Thus, provisions should be made to deal with these issues in the original set of QM plans so that there is a ready mechanism to preserve the integrity of the QM process and, more importantly, retain the high level of trust that is so essential to this project delivery method. This is yet another example of why planning for quality is the major element of DB project success.

Construction Quality Management Planning

The design details define construction quality requirements, and it is the owner's task to allocate QA/QC responsibilities in a manner that suits its risk tolerance and comfort level. Thus, construction QM plans must flow out of the contract, and the contract will embody the requirements outlined in the DB RFP. It would follow that owners who must commit themselves to the cost of construction prior to approving the project's final design, as happens in DB, would devote a significant portion of their DB solicitation packages to defining the required QM process. That, in turn, would cause design-builders to prepare design QM plans which detail their proposed process for each specific project that can be evaluated as a part of the selection process. Additionally, this also ensures that the design-builder has included an appropriate amount for QM in its price proposal. If this did not happen, the owner's only recourse is to require this comprehensive planning to occur post-award.

Two of the authors of this book recently completed a comprehensive study of QA in DB transportation projects for the National Cooperative Highway Research Program (NCHRP) (Gransberg et al. 2007). The study included a content analysis of DB RFPs from projects across the nation and a survey of all DOTs, which means the results of this study provide definitive guidance for construction QM planning. Additionally, the study covered all types of DB projects undertaken by state DOTs, which included both horizontal road and bridge jobs as well as vertical architectural building projects. Since QM is an administrative rather than a technical contract requirement, the majority of the process will be about the same for every owner regardless of project types. As a result, the findings of that study are generally applicable to the entire DB industry.

In trying to identify certain trends about assigned responsibility for conducting the construction QA and QC functions on DB projects, that NCHRP DB QA study looked for specific activities in the solicitation document analysis, such as assigning responsibilities for the following QA and QC activities:

- Performance of shop and/or working drawing and submittal review and approval
- Establishment of horizontal and vertical control on the project site
- Performance of routine QC inspections
- QC testing
- Nonconforming work (punchlist)

Construction Submittal Review

The study completed a content analysis of DB project RFPs to identify how owners assigned the responsibility for construction submittal reviews. The majority (62%) assigned all submittal reviews to the design-builder. This makes sense because the designer-of-record (DOR) works for the design-builder in DB contracts. In 15% of the projects, the owner retained the responsibility and risk of reviewing and approving construction submittals. In the remaining 23% of cases, the design-builder was responsible for reviewing and approving the submittals while the owner retained an active role in verifying the design-builder-performed review. Also confirming this finding was the survey response to the question "Who primarily performs the technical review of construction submittals?" 72% of the respondents indicated that this QA responsibility is placed on the design-builder, while the other 28% indicated that the owner or a third party would review the shop drawings. An RFP issued by the Florida DOT presents a summary of this type of review:

> The Design-Build Firm shall be responsible for the preparation and approval of all Shop Drawings. . . . The Department shall review the Shop Drawing(s) to evaluate compliance with project requirements and provide any findings to the Design-Build Firm. The Department's procedural review of shop drawings is to assure that the Contractor and the EOR [engineer-of-record] have both accepted and signed the drawing, the drawing has been independently reviewed and is in general conformance with the plans. The Department's review is not meant to be a complete and detailed review. (FDOT 2003)

Some owners limited their review to only specific elements of the project that were of specific interest, such as structures or lighting (see MnDOT 2001, SDDOT 2000, and WSDOT 2004). The review in these situations was again just to ensure compliance with contract requirements, and the RFP required that the submittals be first reviewed by the design-builder. This furnishes added confidence in the quality of the constructed product for features of specific interest without usurping the design-builder's responsibility for total QM. It also reduces the amount of construction administration that the owner must conduct to only that which adds value to the process.

Horizontal and Vertical Control

Establishing horizontal and vertical control on the project site is essential to ensuring the overall quality of a project. The NCHRP DB QA study RFP content analysis found that 42 solicitation documents contained specific information on this topic. In all cases, the owner gave control of the construction staking and grades to the design-builder. In 88% of the cases, the design-builder was also responsible for establishing horizontal and vertical control, except for some limited pre-proposal surveying such as that referenced in this Mississippi DOT RFP:

> The Department will establish, one time only, State Plane Coordinate System horizontal control monuments. It shall be the responsibility of the Contractor to establish additional control as may be required to facilitate the staking of the right-of-way. (MDOT 2005)

In the remaining 12% of the projects there was a more extensive shared responsibility between the owner and the design-builder. This also occurred in two examples from the Eastern Federal Lands Highway Division (EFLHD) and the Ohio DOT shown below:

> EFLHD: The Government has established basic survey control points for vertical and horizontal control of the project. (EFLHD 2001)
>
> Ohio DOT: The Department survey crews have provided the necessary survey requirements, listed below: Mainline centerline control and bench marks; Mainline monumentation control . . . Vertical clearances for overhead structures, to serve as a check for the existing vertical clearances. (ODOT 2001)

The survey confirmed the practice of placing the horizontal and vertical control responsibility in the hands of the design-builder, as respondents indicated that this was the case 92% of the time. This is significant in that conducting these tasks carries with it liability for errors. Construction measurements are the fundamental metric that can be checked to ensure that the work complies with the design, because an error in setting the initial reference points from which all measurements are made can carry enormous costs.

Routine Construction Inspection

Routine construction inspection is another construction quality aspect that has overwhelmingly been shifted away from the owner in DB project delivery. An RFP issued by the Virginia DOT (VDOT) clearly delineated this shift: "The construction QA person will perform all of the construction inspection and sampling and testing work that is normally performed by the VDOT" (VDOT 2002). Once again, the content analysis showed that the majority of the projects (82%) assigned routine construction inspection to the design-builder or to both the design-builder and an independent firm. The owner retained the construction inspection duties in only 13% of the cases and shared the duties with the design-builder in 5% of the cases.

There are a number of ways to allocate construction inspection responsibility. The content analysis showed that the owner should assign this responsibility on a project-specific basis rather than merely as a matter of policy where all projects have the same assigned responsibility. Here are examples that show how the North Carolina DOT, an owner with extensive DB experience, dealt with routine construction inspection requirements on three different projects:

1. Design-Builder Responsibility: "The Design Build Team shall provide a schedule indicating the minimum number of inspectors that will be supplied at different stages during the project duration." (NCDOT 2005a)
2. Owner Responsibility: "Construction engineering and management will be the responsibility of the Design-Build Team. Construction

Engineering Inspection [by a third party] will NOT be required in this contract." (NCDOT 2002)

3. Shared Responsibility between Design-Builder and an Independent Firm: "Construction engineering and management, including quality control and inspection, will be the responsibility of the Design Build Team. . . . The Design-Build team shall employ a private engineering firm to perform Construction Inspection for all work required under this contract. This private engineering firm is to be a separate entity, unaffiliated with the Design-Builder in any way. Private engineering firms must be prequalified under the Department's normal prequalification procedures prior to bid submission. This Scope of Work describes and defines requirements for the construction inspection, materials sampling and testing, and technician level contract administration by the private engineering firm (commonly referred to as "Construction Engineering & Inspection" (CEI) firms) required for construction of this project." (NCDOT 2005b)

One can see that this owner has made a conscious decision in each case to apply an appropriate level of control based on the technical requirements of each project. This approach allows the owner to optimize the cost of construction QC based on project characteristics and quality risk, as was shown in Figure 6-3.

The Washington State DOT (WSDOT) furnishes a fourth approach to construction inspection responsibility distribution and provides a clear explanation in their *Design-Build Guidebook*. This owner's approach to routine construction inspection is described as follows:

> WSDOT's inspection involvement will be less extensive than under design-bid-build, depending on the construction schedule and the type of project. The primary role is to monitor the progression of the construction against the Construction Documents submitted by the Design-Builder. *The inspector's authority has not changed, although [the inspector's] work will be coordinated with the Design-Builder inspector.* On projects where WSDOT performed final design on portions of the project, the WSDOT inspector's role will be similar to that under design-bid-build projects. With mixed assignments on-site, the WSDOT and Design-Builder inspectors will need to maintain close coordination to ensure none of the required QC measures are missed. (WSDOT 2004; emphasis added)

Witness and Hold Points

"Witness and hold points" are milestones on the DB project schedule that are designated in advance to permit the owner time to personally inspect specific features of work or processes regardless of who is assigned routine construction inspection duties. WSDOT implemented this policy in their RFP for the Thurston Way Interchange project through the use of this DB construction inspection technique drawn from the building construction industry (Gransberg et al. 2006). A "witness point" is defined in the WSDOT guidebook as an oversight activity, and "hold points" are mandatory inspections held at specific points in construction progress (WSDOT 2004). Often these are points where the design-builder is ready to bury a major feature of work, such as a utility line, or to cover a feature up by casting it in concrete. They represent points in the DB project schedule where the design-builder must notify the owner and arrange for a mutual inspection of those features of work that are identified in the contract. The WSDOT RFP language is:

> Witness and Hold Points are to be established where notification of WSDOT is required for WSDOT's option of observing or visually examining a specific work operation or test. Witness Points are points identified within the inspection plan which require notification of WSDOT. Work may proceed beyond a witness point with or without participation by WSDOT provided proper notification has been given. Hold Points are mandatory verification points identified within the inspection plan beyond which work cannot proceed until mandatory verification is performed and a written release is granted by WSDOT. Witness and Hold Points should be identified in the construction process where critical characteristics are to be measured and maintained, and at points where it is nearly impossible to determine the adequacy of either materials or workmanship once work proceeds past this point. . . . The QC Plan shall contain inspection plans for each construction work item included in the project whether performed by the Design-Builder or a subcontractor or vendor. (WSDOT 2000)

The Arizona DOT also uses witness and hold points as a mechanism to control the adequacy of routine construction inspections (ADOT 2002). In doing so, they set "quality checkpoints" coupled with quality incentives. Typically, ADOT covers DB construction inspection by "requiring the D/B firm to provide independent inspections and materials testing with ADOT oversight and having the contractor perform the QC and the QA by an independent portion of the D/B firm with ADOT providing oversight and independent sampling and testing" (ADOT 2002).

Quality Control Testing

The central aspect of construction quality control is QC testing. This is also where the NCHRP DB content analysis found that owners have overwhelmingly given responsibility to the design-builder. In fact, in the 39 projects that specifically mentioned construction QC testing in the DB RFP, the owner did not retain control in any of them. In all but two cases, the design-builder was assigned direct control. In the two exceptions, a third party was required to perform part of the testing. However, owners normally retain the right to make further inspections or to perform their own verification and acceptance testing. The following example from an RFP issued by the Mississippi DOT illustrates how they gave the responsibility of construction QC testing to the design-builder while retaining verification and acceptance testing responsibilities:

> The CONTRACTOR is required to conduct concrete and asphalt sampling and testing in accordance with MDOT Standard Specifications for Road and Bridge Construction. . . . The CONTRACTOR may elect to conduct other sampling and testing for his own benefit . . . [MDOT] or its duly authorized representative may conduct QA inspections, verification sampling and testing for concrete and hot mix asphalt, all other acceptance testing, and independent assurance testing. (MDOT 2005)

The Texas DOT/Texas Turnpike Authority (TTA) was one of the exceptions in their DB RFP for State Highway 130, a mega-project with its own unique characteristics. Because of the magnitude of the project, TTA chose to rely on an independent QA firm. Its RFP indicated the independent firm's responsibilities as follows:

> For quality assurance purposes, the department shall provide or contract for, independently of the design-build firm, any inspection services or verification testing services necessary for acceptance of the transportation project. (TTA 2001)

These two exceptions follow the idea that the owner must ultimately ensure that the construction QC system supports rather than constrains project progress. By either giving the design-builder full responsibility or by inserting an independent quality firm, the owner is ensuring that it has clearly delineated the requirement for QC testing during the course of construction. Ultimately the owner bears the risk of substandard quality and must be fundamentally comfortable with the QC test plan being able to represent overall project quality.

Nonconforming Work

The final construction QC issue is the disposition of nonconforming work (often called the "punchlist") and determining the party that will be responsible for making the final inspections and reporting final conformance. Both owners and design-builders recognize that a "zero defects/catch-and-punish" approach to construction quality management is both impossible and counterproductive. Therefore, accepting the fact that nonconforming work will occur, the design-builder should develop procedures in its QM plan to identify, evaluate, and correct nonconforming work. The Colorado DOT requires design-builders to cover the following five items in their QM plans:

1. Methods to investigate the cause of systemic nonconforming work and to determine what corrective action is needed to prevent recurrence.
2. Methods to analyze all processes, work operations, quality records, service reports, and owner audits to detect and eliminate the possibility of systemic nonconforming work from occurring.

3. Methods to prioritize corrective and preventive action efforts based upon the level of risk to the quality of the Work.
4. Controls to ensure that effective corrective and preventative actions are taken when the need is identified.
5. Methods to implement and record changes in procedures resulting from corrective and preventive actions. (CDOT 2000b).

Twenty-five of the projects in the NCHRP study specified responsibility for reporting nonconforming work. Intuitively, this is the point in construction QM where most of the owners seem to have retained control. In 64% of the RFPs analyzed, the owner retained the responsibility for reporting and determining nonconforming work. For example, a Utah DOT RFP states: "Nonconforming Work is Work that *the Department determines* does not conform to the requirements of the Contract Documents" (UDOT 2004; emphasis added). This is a clear statement of the owner's perceived responsibility for final acceptance of the completed project.

Interestingly, in 32% of the projects, the design-builder was assigned responsibility for conducting inspections and preparing the reports of the nonconforming work. This is not meant to imply that the owner was absolving itself from responsibility for final acceptance of the project, but, rather, that the owner was requiring the DB team to conduct what would best be termed a "rolling punchlist." This approach reports nonconformance as it is discovered and encourages the design-builder to make corrections as soon as practical rather than waiting until the end of the project. These inspections are used as the basis for a report that documents the findings of these inspections. Presumably the owner would then use this as a basis on which to conduct its own final acceptance activities. The two following examples from Colorado and Utah show how this was expressed in RFPs:

> Colorado: "The Contractor shall establish and maintain a nonconformance system and procedures for uniform reporting, controlling and disposition of nonconformance." (CDOT 2000a)
>
> Utah: "The CPOC [construction proof of compliance] shall identify and document in a nonconformance report (NCR) all elements of the Work that have not, or are believed to have not, been constructed in accordance with the approved drawings and specifications. The NCR shall be submitted to the IQM [independent quality manager] in writing within 24 hours of identification, and a copy sent to the design engineer. . . . The Department will not grant acceptance for any portion of Work that has an outstanding NCR." (UDOT 2004)

A word of caution on this method should be observed. Design-builders may be afraid of reporting nonconforming work. Some industry members feel that their record for nonconforming work may be held against them when they are being scored on the next-best-value design-build project. This attitude violates the spirit of openness and trust that must be established to successfully deliver a DB project. Thus, the owner must make it clear at the pre-work meeting that quality deficiencies identified, reported, and corrected will be viewed as positive for future projects. Additionally, the timely disposition of the project punchlist will also be evaluated and scored accordingly.

Verification and Acceptance Testing

To support final project acceptance and payment, final inspection responsibility as well as any verification or acceptance testing duties must be determined. The NCHRP DB RFP content study analyzed who was responsible for the verification and acceptance testing. As with the report of nonconforming work, owners have generally retained this quality function. In fact, of the 40 projects that listed responsibility for verification and acceptance testing, 88% assigned the responsibility to the owner. Another 5% required an independent firm that worked directly for the owner to accomplish this. Interestingly, three RFPs assigned this responsibility to the design-builder. The survey responses support these findings by indicating that 88% of the time the owner or a third party hired by the owner performed verification and acceptance testing, while the design-builder was responsible only 12% of the time. Although the majority of the owners retained the verification and acceptance testing responsibilities, this does not mean that they performed the actual tests with their own forces. Some owners indicated in the RFP that the owner reserved the right to appoint a representative to perform the tests (see the example from the Florida DOT below). Here are several standard state DOT RFP clauses:

> Arizona: The design-build firm shall be responsible for the quality of the construction and materials incorporated into the project and is responsible for most QC actions. The Department has the responsibility of determining the acceptability of the construction and the materials incorporated into the project. The Department will use the results of the firm's inspection, sampling and testing, and the Department's Surveillance Inspection, and Verification Sampling and Testing to determine the acceptability of completed work items and for final project acceptance. Verification Sampling and Testing will be performed by the Department to validate Design-Builder Sampling and Testing as well as the quality of the material produced. (ADOT 1997)
>
> Florida: The Department or Department's representative will perform independent assurance, verification and resolution testing services in accordance with the latest Specifications. The Design-Build Firm will provide quality control testing in accordance with the latest Specifications. (FDOT 2003)
>
> Minnesota: The Department, through its owner quality assurance (QA), will have the primary responsibility for verification of the quality of both the design and construction work. The Department reserves the right to conduct inspection, sampling, testing, and evaluation associated with QA and IA [independent assurance]. (MnDOT 2001)

Mississippi: [MDOT] or its duly authorized representative may conduct QA inspections, verification sampling and testing for concrete and hot mix asphalt, all other acceptance testing, and independent assurance testing. (MDOT 2005)

Utah: The Department will be responsible for construction QA. The Department will perform the same inspections and tests it performs on a standard design-bid-build project. (UDOT 2005)

Design-Builder Verification and Acceptance Testing

As mentioned above, three of the projects assigned responsibility to the design-builder for the verification and acceptance testing. One of these was for a toll collection and revenue management system where the verification and acceptance testing was based on a 60-day trial period after completion. For projects involving highway construction, however, the two projects that used design-builder verification and acceptance testing were mega-projects: the Colorado South East Corridor (SEC) Multi-Modal Project, and a major urban interstate makeover, I-64 in Missouri. Here are excerpts from the RFPs for both projects:

Colorado: *In cases where inspections are to serve as the basis for compliance verification,* the Contractor shall prepare detailed inspection procedures and submit these to the SEC Representative for review. *The Contractor shall conduct each inspection in accordance with the inspection procedures* reviewed by the SEC Representative; no inspection shall be performed prior to obtaining the SEC Representative review of such inspection procedures. The Contractor in a suitable inspection report clearly showing if the inspection passed or failed based on the "pass/fail criteria" established in the procedure, shall document the results. (CDOT 2000a; emphasis added)

Missouri: The following quality planning aspects shall be included in the Quality Manual: . . . The Quality Assurance staff position *responsible to perform the verification responsibilities including inspection, checking and testing. . . .* The method of performing Quality Assurance verification responsibilities including inspection, checking and testing. (MODOT 2006; emphasis added)

In another mega-project, the San Joaquin Hills Transportation Corridor, an independent firm was to be retained for the acceptance and assurance responsibilities. The RFP states:

The Construction Engineering Manager [employed by an independent firm] shall be responsible for coordinating and directing all Acceptance and Assurance inspections, sampling and testing to he conducted hereunder. (SJHTCA 1991)

Finally, in Utah's I-15 mega-project, a third party was specifically listed as assisting the owner to fulfill its verification and acceptance testing responsibilities (UDOT 1997).

To summarize this section, it is fair to say that in the majority of the cases, the owner assigned the responsibilities for construction QM primarily to the design-builder while retaining either traditional QA or PQA for itself or a third party consultant that worked as the owner's agent.

Post-Award Construction Quality Plans

In the NCHRP study's RFP content analysis, 45% of the projects required some part of a construction quality plan either before or after award of the contract, and 40% required an entire post-award construction quality plan (presumably including both QC and QA) that would not be scored in the proposal evaluation. In addition, 41% of the survey respondents also indicated that they required a post-award construction QM plan. The San Joaquin Hills Transportation Corridor RFP specifically required post-award construction QC and QA plans. In four other projects, only post-award construction QC plans were required.

Most often, the document required the quality plan to be submitted for owner approval within a specific number of days before the start of construction. For example, a project in Washington, D.C. required the contractor to comply with the following: "Before the start of *any construction*, submit a written quality control plan for acceptance" (EFLHD 2001). In another example from the Mississippi DOT, a specific time was listed for the submission of the construction QC plan and the requirements:

> The contractor shall submit a Quality Control Plan that outlines how the contractor shall assure that the materials and Work are in compliance with the contract Documents. The initial plan shall be submitted to the [DOT] for review and approval at least 30 days prior to the beginning of any construction activity. (MDOT 2005)

The New Mexico State Highway and Transportation Department (NMSHTD) included the following in an RFQ about a construction quality plan:

> The Contractor will be required to plan, implement and provide a Quality Assurance/Quality Control (QA/QC) Program for its design and construction operations. . . . The Department will review the Contractor's program to assure that it meets guidelines and minimum requirements established by the Department. Department approval of the program will constitute Department agreement that it meets these criteria, but the Contractor shall maintain ownership of the program and shall be fully responsible for its execution. (NMSHTD 2001)

Some of the post-award construction quality plans were required as a follow-up to a draft or summary presented in the proposal. This was the case in 13 of the 17 projects that required a post-award construction quality plan. Those RFPs first required a pre-award draft of the plan so the owner understood and could evaluate the design-builder's QM approach before awarding the contract. The rationale seems to have been that since the design was not complete, requiring a complete quality plan in the proposal was neither feasible nor necessary. It also served to reduce proposal preparation costs for the competing design-builders. The Minnesota DOT uses this method; one of its RFP states:

> [After award,] the Design-Builder shall submit a Construction Quality Management Plan (CQMP) (based on the Draft CQMP submitted in its Proposal) that addresses Construction Quality Control (CQC), including coordination of the Department's Construction Quality Assurance (CQA) and Independent Assurance (IA) procedures. (MnDOT 2001)

It would seem that this last approach would be the most reasonable. It makes QM planning an evaluated portion of the proposal but does not impose an undue burden on the competitors by asking for a detailed document that is based on their hypothetical approach to the project. Additionally, because the winning proposal typically becomes a part of the contract (Beard et al. 2001), it permits the award to be made without the need to subsequently modify the contract as the QM plans are modified in accordance with the final approved design.

Finally, it is surprising that nearly half the RFPs did not ask that the design-builders furnish some detail as to their QM approach for the project. This group were obviously using the "quality by qualifications" approach (Gransberg and Molenaar 2004) to articulating their requirements for QM, which relies on the evaluation of the design-builder's qualifications and past performance record to ensure that quality will be brought to the design and construction through the caliber of the personnel and firms that complete the work.

Construction Quality Personnel Requirements

The qualifications of the people who perform the construction QM functions on a project drive the reliability of the results of the inspections and tests. Accordingly, the owner should be interested in the people who supervise and perform these inspections and tests. As discussed in Chapter 4, most owners require that a project or construction quality manager be listed in the proposal as part of the competitive evaluation. Also, as was discussed in the same chapter, the primary qualifications for these personnel were experience and professional licensing or certification as appropriate for their assigned duties, as well some education requirements. Table 6-3 shows four possible construction QM schemes that can be used on a typical DB project.

Fifteen of the projects in the NCHRP study had requirements in the RFP for construction quality personnel who must be identified and approved after the contract is awarded. The majority of these projects (73%) required that all personnel that had a construction quality function or task be identified with their levels of

Table 6-3 Construction Quality Management Organizations with Project Quality Assurance[a]

	Construction QA	Construction QC	Construction PQA
Type 1	Design-builder	Design-builder	Owner or 3rd-party consultant[b]
Type 2	Owner or 3rd-party consultant	Design-builder	None
Type 3	Design-builder and Owner	Design-builder	None
Type 4	Owner	Design-builder or 3rd-party consultant	None

[a] Relates to Table 4-3.

[b] 3rd-party consultant is an independent firm retained to conduct QA and QC or independent assurance responsibilities on behalf of the owner.

certification and other qualifications. Many (64%) of them also specified the number of full-time personnel that must be assigned to the project. An excerpt from a South Dakota RFP is a good example:

> The CQMP [construction quality management plan] shall provide the information regarding the Design/Builder's organization in providing quality management of all of the construction processes. The number of full-time equivalent employees with specific Quality Control responsibilities shall be included, as well as a chart showing lines of authority and reporting authority. . . . The Design/Builder shall identify the names, positions, qualifications, duties, responsibilities and authorities of each person proposed in a quality function for construction. (SDDOT 2000)

Another 33% of the RFPs required that the qualifications of QC testing and/or inspection supervisors be submitted and approved after award of the contract. The following comes from the EFLHD RFP describing the contents of a post-award construction QC plan:

> Personnel qualifications[:] Document the name, authority, relevant experience, and qualifications of person with overall responsibility for the inspection system. . . . Document the names, authority, and relevant experience of all personnel directly responsible for inspection and testing. (EFLHD 2001)

A quality product requires people who perform in a qualified manner. To ensure that construction is completed at the required level of quality, the owner must know the qualifications of those who will work on its project. QM plans cannot be effectively implemented by marginally qualified personnel.

As can be seen from the various examples and statistics cited in this chapter, many methods and approaches can be used to attain the required construction quality on a DB project. The main conclusion is that each project is unique; to gain the greatest benefit of DB, each project must have a construction QM plan crafted specifically for that project. As owners gain more experience with DB, they will better understand how best to manage construction quality in a DB project, and as design-builders craft the QM plans that nest within their market's owner plans, the general form and format of construction QM will eventually become as institutionalized as the DBB QA/QC programs are today.

Three rules must be followed to assure quality of the constructed project:

1. The owner must clearly articulate the construction QA roles and responsibilities in its RFP so the design-builder can estimate the resources necessary to satisfy that requirement.
2. Each DB project is unique and, as such, each project's QM plans should be customized for its environment, technical complexity, and other pertinent issues that affect construction quality.
3. The plans for construction QM must flow out of and be compatible with the design QM plans to furnish continuity of the entire quality life cycle.

References

Abbott, L. (1955). *Quality and competition: An essay in economic theory.* Columbia University Press, New York.

American Society for Quality (ASQ). (2008). "Quality glossary." *American Society for Quality,* <http://www.asq.org/info/glossary/> [Dec. 26, 2008].

Arizona Department of Transportation (ADOT). (2002). "Design-build procurement and administration guide." *ADOT Construction Group, Intermodal Transportation Division,* <www.azdot.gov/Highways/ConstGrp/Value_Quality/PDF/DesignBuildGuide.pdf> [March 12, 2010].

ADOT. (1997). *Contracts and specifications section and engineering consultants section, design build package request for statements of qualifications and requests for proposals: Cortaro Road interchange MP 245.8,* Pima County, Arizona Department of Transportation, Phoenix, Ariz.

Beard, J. L., Loulakis, M. C., and Wundram, E. C. (2001). *Design build planning through development.* McGraw-Hill, New York.

Broh, R. A. (1982). *Managing quality for higher profits.* McGraw-Hill, New York.

Clark, T. (2007). "Developing operating procedures for projects involving multiple organizations using a linear responsibility chart." *Packaging Digest,* 12, 3.

Colorado Department of Transportation (CDOT). (2000a). *Request for proposal, Book 1, Southeast Corridor multi-modal project,* RFP No. 11584, Colorado Department of Transportation, Denver, Colo.

CDOT. (2000b). *Request for proposal, Book 2, Southeast Corridor multi-modal project,* RFP No. 11584, Colorado Department of Transportation, Denver, Colo.

Dortman, R., and Steiner, P. O. (1954). "Optimal advertising and optimal quality." *Am. Econ. Rev.,* 5, 831.

Eastern Federal Lands Highway Division (EFLHD). (2001). *Request for proposals IBC-8888(012),* Book 1, Federal-Aid Project No. IBC-8888(012), District of Columbia Department of Public Works, Washington, D.C.

Edwards, C. D. (1968). "The meaning of quality." *Quality Prog.*, 1, 36–39.

Florida Department of Transportation (FDOT). (2003). *Request for proposal, milling and resurfacing of Interstate 95,* Financial Projects No. 409846-1-52-01; Federal-Aid Project No. 0953 102 I; Contract No. E-5F84, District 5, Florida Department of Transportation, Tallahassee, Fla.

Gilmore, H. L. (1974). "Product conformance cost." *Quality Prog.*, 7(5), 16–19.

Gransberg, D. D., and Molenaar, K. R. (2004). "Analysis of owner's design and construction quality management approaches in design/build projects." *ASCE J. Mgmt. Eng.*, 20(4), 162–169.

Gransberg, D. D., Molenaar, K. R., and Datin, J. N. (2007). "Quality assurance in design-build projects." *NCHRP Synthesis 38-01,* Transportation Research Board of the National Academies, Washington, D.C.

Gransberg, D. D, Koch, J. E., and Molenaar, K. R. (2006) *Preparing for design-build projects: A primer for owners, engineers, and contractors,* ASCE Press, Reston, Va.

Juran, J. M., ed., *Quality control handbook,* 4th ed., McGraw-Hill, New York.

Keuhn, A. A., and Day, R. L. (1962). "Strategy of product quality." *Harvard Bus. Rev.*, 40(6), 100–110.

Leifler, K. B. (1982). "Ambiguous changes in product quality." *Am. Econ. Rev.*, 72(5), 956–967.

Maynes, E. S. (1976). "The concept and measurement of product quality." *Household production and consumption,* N. E. Terleckyj, ed., National Bureau of Economic Research, Inc., Cambridge, Mass.

Minnesota Department of Transportation (MnDOT). (2001). *Request for qualifications T.H. 52 design-build project S.P. 5502-67, 5507-45, 5508-78; HPPMN45(101), NH014(308), District Six–Rochester,* Minnesota Department of Transportation, St. Paul, Minn.

Mississippi Department of Transportation (MDOT). (2005). *Request for proposals, Addendum 1, a design-build project bridge replacement on US 90 Biloxi to Ocean Springs Bridge, Jackson and Harrison Counties, Mississippi,* Contract No. ER/BR-0003-01(099) 104556/301000–Biloxi/Ocean Springs Bridge, Mississippi Department of Transportation, Jackson, Miss.

Missouri Department of Transportation (MoDOT). (2006). *Final request for proposals for the new I-64 design-build project,* Project No. J6I0978, Book 1, Missouri Department of Transportation, Chesterfield, Mo.

New Mexico State Highway and Transportation Department (NMSHTD). (2001). *Request for proposals, US70 Hondo Valley,* AC-MIP-070-41(35) 264, New Mexico State Highway and Transportation Department, Santa Fe, N.M.

North Carolina Department of Transportation (NCDOT). (2005a). *Design-build package: I-40 from west of SR 1224 (Monte Vista Road) to I-240 in Asheville,* Contract ID: C 201277, Buncombe County, North Carolina Department of Transportation, Raleigh, N.C.

North Carolina Department of Transportation (NCDOT). (2005b). *Design-build package: I-77 from I-485 to SR 2136 (Gilead Road) exit 23,* Contract ID: C 201288, Mecklenberg County, North Carolina Department of Transportation, Raleigh, N.C.

North Carolina Department of Transportation (NCDOT). (2002). *Design-build package: US64 Knightdale bypass,* Project No. 8.14021002; Federal-Aid No. NHF-64(73), North Carolina Department of Transportation, Raleigh, N.C.

Ohio Department of Transportation (ODOT). (2001). *Final report: Six-state survey of construction administration practices and procedures,* Ohio Department of Transportation, Columbus, Ohio.

Persig, R. R. (1974). *Zen and the art of motorcycle maintenance: An inquiry into values*, William Morrow, New York.

Pheng, L. S., and Yeap, L. (2001). "Quality function deployment in design/build projects." *ASCE J. Arch. Eng.*, 7(2), 30.

San Joaquin Hills Transportation Corridor Agency (SJHTCA). (1991). *Design/build services: San Joaquin Hills Transportation Corridor entire corridor*, IFB No. S90-19, Transportation Corridor Agency, Costa Mesa, Calif.

Songer, A. D., and Molenaar, K. R. (1996). "Selecting design-build: Private and public sector owner attitudes." *ASCE J. Eng. Mgmt*, 12(6), 47–53.

South Dakota Department of Transportation (SDDOT). (2000). *Request for proposals, Interstate 229*, Project No. IM229-2(50)2; PCEMS No. 1231, South Dakota Department of Transportation, Pierre, S.D.

Texas Turnpike Authority (TTA). (2001). *Request for proposals to construct, maintain and repair the SH 130 turnpike through an exclusive development agreement*, Texas Turnpike Authority, Texas Department of Transportation, Austin, Tex.

Transportation Research Board (TRB). (2005). *Glossary of highway quality assurance terms: Transportation Research Circular E-C074*, Transportation Research Board of the National Academies, Washington, D.C.

Tuchman, B. W. (1980). "The decline of quality," *New York Times Magazine*, Nov. 2, 38–41.

Tulsa World. (2008). "SemGroup goes bankrupt." *World Publishing Company*, <http://www.tulsaworld.com/webextra/content/2008/SemGroup/default.aspx> [Dec. 26, 2008].

Utah Department of Transportation (UDOT). (2005). *Request for proposals, SR-92; east of I-15 in Lehi, Utah County*, Project No. STP-0092(5)1, Utah Department of Transportation, Orem, Utah.

UDOT. (2004). *Request for proposals SR-265–University Parkway at UVSC*, Project No. NH-0265(5)2, Utah Department of Transportation, Orem, Utah.

UDOT. (1997). *Request for proposals, I-15 corridor reconstruction project*, Project No. *SP-15-7(135)296, Utah Department of Transportation, Orem, Utah.

Virginia Department of Transportation (VDOT). (2002). *Request for proposals, a design-build project design and construction of a safety rest and information center on I-64 Eastbound in New Kent County*, State Project No. 0064-063-F08, PE102, L802, New Kent County, Virginia Department of Transportation, Richmond Va.

Washington State Department of Transportation (WSDOT). (2004). *Guidebook for design-build highway project development*, Olympia, Wash.

WSDOT. (2000). *Request for proposals, Thurston Way Interchange, SR500 MP3.15 to MP 4.73*, Clark County, Washington State Department of Transportation, Olympia, Wash.

SEVEN

Design-Build Change Orders

An old adage says that change is inevitable, and nowhere is this more true than in a typical construction project. The sheer complexity of many projects render it virtually impossible to cover every technical eventuality during the design process. In the design-bid-build (DBB) world, every inaccuracy in the construction documents represents a potential change order with concurrent cost and time growth due to the resulting additional scope. If the inaccuracy is not discovered until late in the process, the cost growth can potentially be enormous, and most changes in scope also come with extensions in project schedules. Therefore, controlling change after contract award is always in the best interest of the project's owner. One of the benefits of design-build (DB) project delivery is its ability to create a single point of responsibility for both design and construction, which effectively minimizes the potential for change orders due to design errors and omissions (Beard et al. 2001).

In fact, research comparing DB project performance to the performance of similar DBB projects has verified this assertion and the ability of the owner to actually accrue the benefits of reduced cost and time growth. Table 7-1 is a summary of research on this particular subject. One can see that owners who select DB project delivery can do so with an expectation that both cost and time certainty will be improved (Warne 2005).

Additional benefits can be accrued by shifting the responsibility for coordinating the design with the construction, as illustrated in the following quotation from a study of a large state university's DB program:

> Owners will realize cost benefits with design-build approaches by converting the project coordination function from a cost management item controlled by the Owner to a profit management incentive for the design-builder. (Reed 2003)

Hence, this leads to an expectation that there will be fewer change orders as well. In fact, one study in the mechanical contracting sector found this to be true, reporting that DB projects had an "87% decrease in the average number of unforeseen

Table 7-1 Summary of Design-Build versus Design-Bid-Build Project Performance Research

Literature Source	Project Type	DB vs. DBB Cost Growth	DB vs. DBB Time Growth	Remarks
Allen 2001	Buildings	18% less	60% less	110 Navy military construction projects
Konchar and Sanvido 1998	Buildings	4.5% less	4.5% less	~350 projects
Bennett et al. 1996	Buildings	4.5% less	4.5% less	~250 projects in the UK
Gransberg et al. 2003	Buildings	16.4% less	19% less	88 federal buildings
Riley et al. 2005	Mechanical	71% less	NA	120 mechanical projects
FHWA 2006	Transportation	5.4% less	18%–71% less	304 transportation projects
Warne 2005	Transportation	1%–6% less	76% of DB projects completed early	21 transportation projects
Hale et al. 2008	Buildings	2% less	50% less	77 Navy military construction projects

change orders" and "the average size of unforeseen change orders was 86% smaller" (Riley et al. 2005). Another study in transportation found that "The subset of design-build projects had fewer change orders than the comparable design-bid-build projects" (FHWA 2006). So again the expectation appears to be confirmed in the literature. However, one must remember that cost growth is also generated from claims as well as change orders. It is important to keep the two causes of cost growth separate when making professional judgments regarding project performance.

A classic textbook on construction contracting defines a change order as:

> Change orders document changes from the original scope of the contract, confirm schedule revisions and set forth other modifications. (Barrie and Paulson 1991)

Thus, change orders essentially relate to modifications in the contractual scope of work and the contractual time period for completing the scope. They also relate to purely legal changes to the contract language which generally do not cause cost or time growth in the project. This definition applies to all types of project delivery methods for both design and construction. A "claim," on the other hand, is essentially a notification to the owner that the other party believes the scope of the original contract has changed. If the owner agrees, a change order is negotiated and incorporated into the contract.

If the owner does not agree that the contract scope has changed, then both parties turn to the contract's disputes clause and begin the process of attempting to resolve the dispute. If at any point in the process both parties reach an agreement, the claim can either be dropped or converted to a change order by bilateral agreement. If the process continues to its final extent (i.e., dispute resolution by some legally empowered body such as an arbitration panel or the courts), the final judgment will be rendered. If it is against the owner, cost growth due to the claim will be incurred in the project. Additionally, both parties to the dispute often incur additional costs to promulgate the dispute resolution process in terms of legal fees, internal administrative costs, and reduced profits. All these create project-related cost growth, regardless of whether they are formally applied to the final contract amount. Thus, avoiding claims should be the primary objective of any DB contract's changes clause.

Again, the literature shows that there is an expectation by owners that implementing DB will reduce claims (Songer and Molenaar 1996). The FHWA's *Design-Build Effectiveness Study* (2006) found that "claims represented less than one-tenth of one-percent of total [DB] project costs." Konchar's and Sanvido's seminal study (1998) on project delivery method project performance indicated that because DB allows the design-builder to control the details of design, it creates an environment that promotes cost control. Another study discussed the motivations involved in creating a bias *toward* cost and time growth in DBB project delivery when it stated:

> In DBB the contractor is tempted to under-price the basic contract in order to get the award and then try to make up profit margins by generating change orders and claims. (Runde and Sunayama 1999, pp. 46–48)

An audit of California State University's DB process confirms that DB changes this motivation through application of the theory of single point of responsibility, as follows:

> Other factors influencing the desire for single point responsibility include forces such as partnering, total quality management, team building, and alternative dispute resolution. The design-build entity is responsible for the adequacy of design and any construction defects. As a result, the owner avoids these types of claims, which are common with the traditional delivery system. (Reed 2003)

Given the above discussion, one can conclude that the owner can look forward to a project with fewer legal hassles if the design and construction can be administered in a manner that promotes the benefits of DB project delivery. Chapter 4 discussed the benefits of formally partnering DB contracts and cited research that showed that partnering greatly reduced the incidence of claims (Gransberg et al. 1999). Therefore, the remainder of this chapter will be devoted to discussion of changes that occur in scope, time, and other common issues such as *force majeure* which are not subject to dispute and hence are not considered claims.

Design-Build Change Order Entitlement

Change orders are the right and privilege of the owner. It is, after all, the owner's project being designed and built with the owner's money. Therefore, the owner can always change its requirements if it is willing to pay for that privilege after contract award. Consummating a change order in DB involves quantifying the change's impact on the project's schedule and budget, as well as furnishing the legal authorization to make the change. In DB a number of issues make quantifying these impacts less straightforward than they are in DBB. The major issue revolves around the fact that the design is not 100% complete when the contract is awarded. Therefore, the technical baseline may be hard to measure. The owner will express its design intent for the project in performance terms, using both design criteria and performance specifications in the RFP's scope of work. The design-builder then takes the owner's scope and further refines it by expressing the design assumptions it made to arrive at its price proposal. Thus, when the contract is awarded, there will be plenty of opportunities for each party to have misinterpreted the other's intent.

Next, because the primary reason why owners choose DB project delivery is to compress the schedule (Songer and Molenaar 1996), most DB projects are awarded with an aggressive schedule; this may include starting construction before 100% design completion, making the need to review and approve design submittals a critical path activity. Therefore, a well-defined process for resolving the professional differences of opinion that are inevitable in this project delivery system during design reviews is mandatory. Usually, these are dealt with by the owner deciding whether the conflicting interpretation of a given performance requirement is adequate. If the owner is not willing to accept the design-builder's design solution or construction method, it may then find itself in a position of having to pay for a change order if the contract does not support the owner's chosen interpretation of a given criterion.

Types of Design-Build Changes

In DBB projects, there are typically three major reasons why a construction contractor would be entitled to a compensable change order. These are:

- Owner-generated impacts on construction: technical scope changes, owner interference with progress, or disruption caused by the owner or other parties for whom owner is responsible under the terms of the contract.
- Unforeseen conditions: previously undiscovered subsurface obstacles or conditions, *force majeure*, and changes in the regulatory or legal preconditions that relate to the project.
- Design issues: design errors and omissions, as well as ambiguities in the construction documents.

In spite of some contrary assertions in the popular literature, from the owner's perspective implementing DB does not eliminate change orders. DB projects will still be carried out in locations where there are unforeseen site conditions. They will also remain susceptible to the whims of the political process and have to be brought into compliance with shifting environmental regulations, and it goes without saying that no delivery method can shield an owner from acts of God. One author puts it this way:

> In design-build projects, the first two categories [shown above] may still result in change orders, but the third category ordinarily does not. Since the contractor [i.e., the design-builder] is responsible for the plans and specifications, it cannot make use of errors and ambiguities in them to claim entitlement to an extra. However, if the owner's criteria from which the contractor prepares the design have ambiguities or omissions, the need to change the design to accommodate new or clarified owner's criteria may result in change orders. (Friedlander 2007)

Merit versus Quantum

Thus, as usual, one must look to the contract to determine whether a design-builder is entitled to a compensable change order. As shown in Chapter 1, Figs. 1-2 and 1-3, the construction documents are no longer a part of the contract. They are a deliverable that flows from the contract's scope of work. Therefore, the caveat that Friedlander added regarding design issues leads the parties to a DB contract to look at the contents of the owner's RFP to determine whether a change order is appropriate. Often the contract will contain the owner's performance criteria, which are discussed in detail in our first book on this topic (Gransberg et al. 2006). Performance criteria are, by definition, open to interpretation. Thus, owners must be careful to not impute their personal interpretation of these criteria if they are used in a DB RFP. As stated in our previous book, if there is only one technically acceptable design solution for a given feature of work, then the owner needs to prescriptively specify it in the RFP (Gransberg et al. 2006). If the owner fails to do this and believes that no other solution will be acceptable, then that owner must be prepared to issue a compensable change order to the design-builder to modify that clause in the contract from a performance criterion to a prescriptive specification. Thus, determining entitlement is fairly direct, but determining the cost and time impact of the modification is not nearly as clear. In legal terms, the last example would clearly find that there is "merit" to the design-builder's request for a change order, but determining the "quantum" due for changing from an open-ended performance criterion to a prescriptively specified design solution will be murky if the owner's change order process has not been well defined prior to the advertising and award of the DB contract.

The quantum portion of the change order will be affected by a number of issues. First and probably foremost is the amount of design effort that has been expended on the feature of work that will be modified. If the change occurs early, the cost and time impact may be minimal. For example, if during the initial design meeting an owner decides to alter the design-builder's proposed floor plan to accommodate a change in function not known before contract award, it will probably be able to do so without cost if the overall size of the building is not changed. On the other hand, requesting the same change after the architectural design has been approved and the engineering design is underway will result in lost design effort as well as additional redesign costs and potentially a change to the amount of construction material used. It might also cause a delay in the overall schedule as the affected portions of the building are redesigned.

Design-Build Change Management

Thus, it is vital that the owner and the design-builder understand the DB project's timeline and be particularly sensitive to those periods of time in which the project is most susceptible to changes. Assuming that a typical DB project will have an overlap of design effort with construction work makes this effort particularly important. This discussion should be an agenda item in the DB project pre-work meeting. The pre-work meeting is typically devoted to arriving at a thorough understanding of how the owner and design-builder will do business, and "change management" is an important topic to cover during this meeting. An insightful article on this subject was written by Bruce Hallock of the Nielsen-Wurster Group (2006), and contained the following checklist for the business process of managing change in DB projects:

- Validate the change against the baseline project scope, cost and schedule;
- Define a business case for the change and source and/or method of funding;
- Provide a formal justification for the change;
- Determine the cost and time to implement the change;
- Settle, approve, properly authorize and execute the change order; and
- Communicate the change status to the project, and track the performance of the changed work.

Change Timeline in Design-Build

Figure 7-1 shows the concept of how design can overlap construction in a DB project. It attempts to relate the DB process to the standard design process used in a typical building project. An engineered project such as a water treatment plant or a transportation project will follow much the same process. The reader will merely need to substitute the proper design milestone (e.g., 30%, 60%) as shown below the design phase line in the figure.

Bruce Hallock describes change management as follows:

> Managing change means implementing changes in a planned and systematic fashion. The aim is to more effectively implement new or revised work on an ongoing project, and organization. Changes not only affect the changed

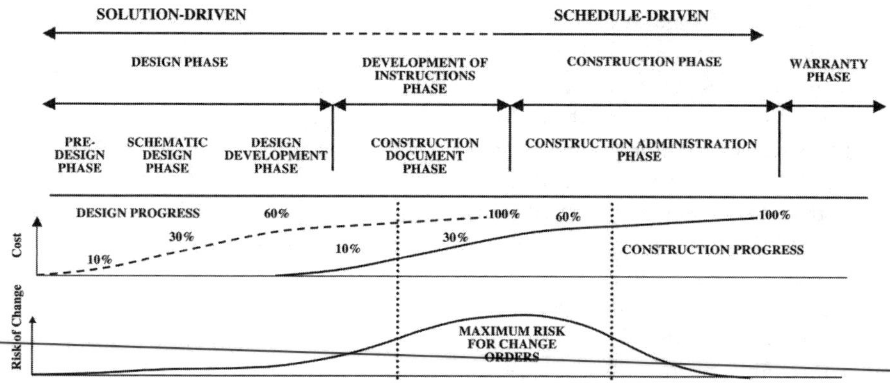

Figure 7-1 DB time line and impact of the risk of change orders.

> work, but when not properly controlled seriously affect the "unchanged" work. Managing changes, whether internally or externally initiated, anticipated or emergent, lie within and are controlled by the project organization. (Hallock 2006)

Figure 7-1 furnishes a model for understanding the "planned and systematic fashion" in which the change must be managed. The project organization identified in the above quote consists of the owner and the design-builder, that is, the parties to the DB contract.

Looking at the figure, one can see that there comes a point where the design-builder must complete various aspects of the design and commit to building them as shown. These design commitments occur in conjunction with the design submittals required by the contract. While the figure shows the classic design phases to ensure that the reader understands what it is graphically portraying, the owner can easily structure the design submittals to mirror the order in which the major features of work within the project are going to be built. This theory and its benefits are discussed in Chapter 3 of this book. The major benefit is that it directly links the design progress with construction progress, allowing the owner to release each design product for construction as soon as it is ready. This also facilitates the change management process by furnishing a much clearer means to "validate the change against the baseline project scope, cost and schedule" (Hallock 2006). The approved design work package becomes the baseline against which the change is validated.

Figure 7-1 also shows that the portion of the DB timeline that is most susceptible to change occurs during the final stages of design and in the middle stages of construction. This is because final design decisions made by various disciplines are being brought together into the same document, and design product is being deconflicted between disciplines in accordance with good design QA practice. Also, the design-builder's cost engineering team will be reviewing the final designs to ensure that they conform to the contract budget and time requirements, potentially necessitating a change to bring the project back within budget or schedule. Finally, construction documents are being released to the construction trade subcontractors who are then reviewing them and finding the conflicts, errors, and omissions missed by the design QA process. Most of these changes occur internally to the design-builder's team and, unless the change is due to owner-generated errors or omissions, will not result in a contractual change order. Nevertheless, it is also in this stage that the owner gets its "last bite at the apple," making final design reviews which potentially uncover elements of scope that are not included in the contractual scope of work.

It is in the best interests of both parties to recognize this critical period for change management and to agree to assist each other in satisfying the demands for each by facilitating the changes that arise as a result of the above cited actions. For instance, if the design-builder's cost engineering review of a feature of work—whose scope was defined by performance criteria—finds that the selected design solution will exceed the budget assumed in the price proposal, and there is a different design solution available that satisfies both the technical and the financial constraints, the owner can facilitate this by allowing the design to change even though it may have previously approved the original design. The design-builder must still bear the burden of the cost of redesign, but it is now in a position to mitigate the impact of the error on its part. Conversely, an owner who finds that it must unexpectedly add a feature of work to the project can ask the design-builder to facilitate this change by offering alternatives to minimize the cost and schedule impact. In fact, a design-builder (with an eye to future work with the same owner) may choose to scrub its budget and see whether there are sufficient unused contingency funds to make the change without additional cost, as a means of strengthening its reputation with this client.

Change Management Best Practices

This discussion leads to the development of a set of best practices for change management in the DB contract. These practices are established on the idea that change will occur despite the best effort of all parties. The Construction Industry Institute (CII) published a report in 1994 that detailed a set of change management best practices (CII 1994). A later study of more recent data confirmed these practices and posed them as 14 questions that an owner and a design-builder can ask themselves when establishing a given project's change management process. Those questions are:

1. Was a formal documented change management process familiar to the principal project participants used to actively manage change on this project?
2. Was a baseline project scope established early in the project and frozen, with changes managed against this base?
3. Were design "freezes" established and communicated once designs were complete?

4. Were areas susceptible to change identified and evaluated for risk during review of the project design basis?
5. Were changes on this project evaluated against the business drivers and success criteria for the project?
6. Were all changes required to go through a formal change justification procedure?
7. Was authorization for change mandatory before implementation?
8. Was a system in place to ensure timely communication of change information to the proper disciplines and project participants?
9. Did project personnel take proactive measures to promptly settle, authorize, and execute change orders on this project?
10. Did the project contract address criteria for classifying change, personnel authorized to request and approve change, and the basis for adjusting the contract?
11. Was a tolerance level for changes established and communicated to all project participants?
12. Were all changes processed through one owner representative?
13. At project close-out, was an evaluation made of changes and their impact on the project cost and schedule performance for future use as lessons learned?
14. Was the project organized in a Work Breakdown Structure (WBS) format and quantities assigned to each WBS for control purposes prior to total project budget authorization? (Zou and Lee 2006)

While the above set of best practices was not developed specifically for DB, they are fully applicable when set into the DB context. For instance the "design freeze" referred to in practice No. 3 relates directly to the design-builder's design commitment in DB. This is a point where the designer-of-record (DOR) has approved the element of work for construction and it is ready for the owner's final review and approval. Essentially, the design-builder is stating that it will "freeze" that design package and will build it according to the design shown in the design submittal. The owner can then use those documents to inspect the final construction and expect to see that feature constructed as designed. Practice No. 4 relates to the previous discussion of the DB timeline where the owner and the design-builder identify those areas which are most susceptible to change and proactively manage those areas as outlined in Practice No. 9. Practice No. 14 speaks to the idea that the design and construction effort should be organized around work packages that spring from the project's WBS, and supports the assertion that change can best be managed during design if the project is organized around features of work that can be designed, reviewed, approved, and constructed in that order, rather than relying on the DBB-oriented design development process of design milestones.

Practice No. 1 speaks to the need to have a well defined change management process in place at project inception. It states that the process should be both "formal" and "documented." The best way for this to happen is for it to be included in the contract and then augmented and expanded post-award to meet the needs of the project. Finally, it says that the process should be "familiar" to all project participants. To accomplish this, the change management process should be covered in detail during the DB pre-work conference to ensure that not only the design-builder but also the owner's project team understand the process and how it will be applied before the need to use it arises.

Design-Build Change Order Process during Design

Remembering that the design is a contract deliverable in DB creates a situation where any alteration to the design-builder's design could potentially affect construction costs. Asking the DOR to move a line in a design drawing is no longer merely shifting some ink around on a big piece of paper. That line could represent a salient design assumption that was used to arrive at the DB project's firm, fixed price. Thus, the change management process during the design phase of a DB project must be radically different from the one used by an owner in a traditional DBB project. This difference stems from the following reasons:

- The sum of the technical scope of work described in the RFP as modified by the design-builder's proposal is the contract scope of work. This scope will be quantified by performance criteria, design criteria, prescriptive specifications, and/or design drawings that can range from conceptual documents to standard owner-required details. Thus, the design change management process must be robust and dynamic enough to account for this wide range of design information.
- The design-builder will have made certain design assumptions in order to furnish a lump-sum price proposal. Unless these assumptions are contained in the proposal or directed by the RFP instructions, they will not be specifically shown in the above-listed contract technical scope of work. Therefore, the design change management process must have a mechanism that provides for the analysis of design-builder design assumptions that are not articulated in the proposal against RFP performance or design criteria. This is to ensure that if the requested change affects one of those assumptions, the design-builder is allowed to correct the price proposal accordingly.
- Because DB projects are normally awarded with an aggressive schedule, design-builders are normally allowed to begin logistics activities as early as is reasonable. Thus, the design change management process must consider the potential impact on the logistics associated with the design change and be prepared to compensate the design-builder if it has to cancel materials orders or trade subcontractors in order to accommodate the change.
- Design changes can include different interpretations of codes and standards that are incorporated by reference in the contract. The design-builder will have included code or standard interpretations in the design

assumptions used to prepare its price proposal. If a code or permitting agency substantially changes the historical interpretation of a given standard, this will constitute a change to the design-builder's design assumptions and should properly be treated as a constructive change to the given code requirement. An example is a seismic code that historically allowed more than one option for a given technical feature, but has been reinterpreted by the code reviewers to require a specific design for the same feature. If the first instance of this reinterpretation occurs as part of the review of a design submittal on a DB project, this would constitute a change to the design-builder's proposed design because neither the owner nor the design-builder could have known that the code agency was going to become more restrictive in its interpretations.

The above bullets encapsulate the issues associated with design risk during DB contract administration. Owners would do well to consider these issues when preparing the RFP and ensure that competing design-builders understand the details of how the design change management process will operate. If the details are unclear at the time of proposal preparation, the design-builders will have no choice but to increase their contingencies to account for possible effects on the project. This could potentially take the price proposals over the owner's budget, making it difficult, if not impossible, to award the contract in a timely manner.

Finally, the design change order process should have two facets:

- Changes to the design that occur *before* the start of construction on that feature of work; and
- Changes to the design that occur *after* the start of construction on that feature of work.

As previously stated, owners often select DB project delivery as a means to accelerate the project's delivery date. Thus, there is a probability that construction activities may have started at the time the need to change the design is identified. Thus, the design change management process should first differentiate between the two states before moving into the development of time, cost, and impact data.

Design Changes before Construction Starts

Changes to the design that occur before the start of construction are fairly straightforward, and an owner can use its DBB process as the basis for developing its DB process. The major difference springs from the fact that in a DB contract design-builders can initiate activities that are typically considered construction activities in DBB at an earlier point in time. An example of such an activity is to lock-in material pricing for large-volume materials such as concrete. A design-builder usually only needs to know the order of magnitude of the amount it needs for a given commodity to be able to lock-in a unit price from a material supplier. Hence, it is always looking to reduce material pricing risk by consummating material supply contracts as early as possible. For instance, if a design-builder's proposal contemplates using roughly 10,000 cubic yards of concrete on a DB bridge project and it feels that the design for those features that require concrete is fairly stable, it does not need to wait until the design is finished to know exactly how much concrete is required before it can sign a contract to furnish at least 10,000 cubic yards of concrete at a fixed unit price per cubic yard. However, if at some point in the design process the owner or other entity beyond the design-builder's control decides to change the main structural system from cast-in-place concrete to structural steel, then the quantity of concrete required would drop substantially and the design-builder would need to renegotiate the material supply contract at a higher unit price. Thus, because the original proposal was based on cast-in-place concrete, consideration would need to given to the cost increase accrued to those cast-in-place concrete features that are not affected by the design change.

Given the above discussion, Fig. 7-2 is a flow chart that describes the design change order process where the change is identified before physical construction has been initiated. From the figure one can see that the first step is to evaluate the change in the context of the RFP performance criteria and the design-builder's DOR should then generate options that are acceptable within the constraints described by the RFP criteria. Once one or more acceptable options are developed, the design-builder must review those alternatives to determine whether they can be accomplished within the existing budget. If additional cost is necessary, the design-builder's team should initiate a value engineering (VE) process to determine whether the options can be modified to make them fit the budget.

If there is an option that satisfies budget constraints, it next needs to be checked to determine whether it can be implemented without extending the schedule. If not, then the design-builder will again initiate a VE process to look for variations that meet schedule constraints. If VE is successful or if the change can be made without budget or schedule impact, the design-builder then prepares a change order proposal for technical review by the owner. If the VE is unsuccessful, the design-builder then prepares a change order proposal including the cost, schedule, and impact of the change for technical, budgetary, and schedule review by the owner. Negotiations ensue and, if an agreement is reached, the owner approves the change order proposal. Then the design, budget, and schedule are updated to reflect the change. If negotiations fail to achieve a result that the owner will approve, the initiator of the change request can either choose to drop the change order request and proceed as previously planned, or allow the design-builder to explore other options if possible, or initiate the dispute resolution process that is prescribed by the contract.

Ultimately, the change will be implemented in some form if it is indeed a technical necessity, and the technical scope of work, as well as the project contract amount and schedule, will be modified to reflect the outcome of the change order management process. At this point it is extremely important that the design-builder notify all the members of its design and construction teams that are affected by the change.

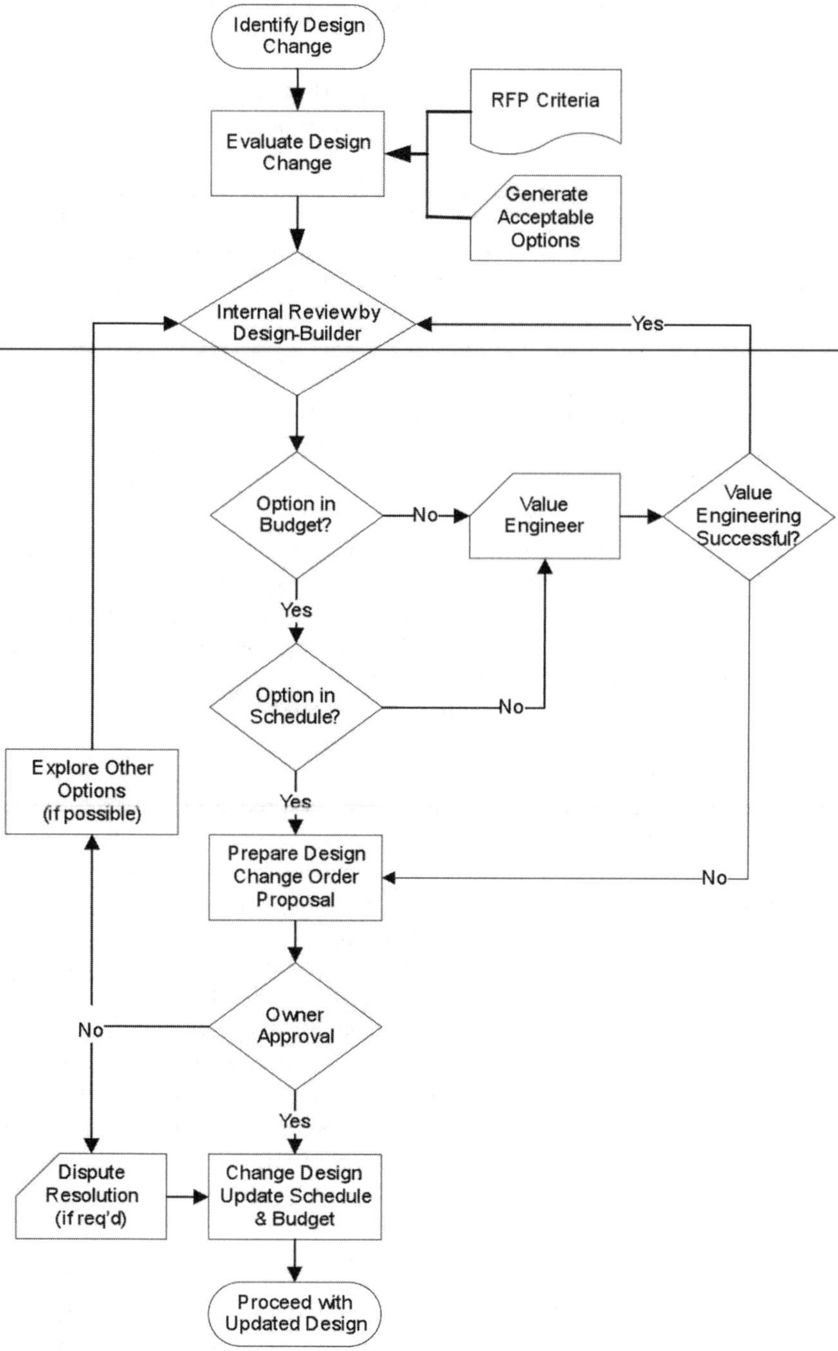

Figure 7-2 DB design change flow chart during design: no construction has started when change is identified.

Design Changes after Construction Starts but before Final Design Is Complete

Because DB allows construction to begin at its earliest possible time, it is probable that a design change could be identified after construction has begun but before the final design is released for construction. This is the point shown in Fig. 7-1 where the maximum risk for change orders begins. The major risk here is that construction will have started on a feature of work whose design must be changed, hence requiring rework. If this is not the case, then the process described in the previous section and shown in Fig. 7-2 applies. There are four situations where these types of changes occur:

- Owner-directed change
- Change mandated by new legislation or regulation
- Change mandated by adverse code interpretation
- Change required by the design-builder, proceeding at its own risk

The first category is probably the most common type of change in a DB project. The pace at which DB projects are often procured can leave little time for the owner to do comprehensive, detailed project scoping like it can do in DBB projects. This is especially true if DB is being used to deliver a project under emergency conditions, such as have occurred in natural disasters or major structural failures. As a result, the owner may find itself needing to add scope to the project after the contract is awarded. When these changes occur after construction has begun, they can become quite expensive and sometimes time-consuming. The major issue in this situation is ensuring that the change is indeed required. Some owner-directed changes are really owner preferences that were not adequately described in the technical scope of work. These often pop up during the design review process when the owner's representatives realize that the design-builder is not intending to build what they expected or perhaps hoped for. Therefore, the first action before instituting the change process is to verify that the change is truly a necessity. If this is found to be true, then the status of both the design and the ongoing construction must be quickly reviewed to mitigate any potential increased costs.

In this situation it is incumbent upon the design-builder to ensure that notice of the impending change gets down to all the design subconsultants and trade subcontractors, with instructions to carefully review their progress and determine whether they can assist the owner in keeping the cost of the change as low as possible by ceasing to work on those features that will be affected by the change. The first risk in this situation is the cost of lost design effort by the DOR and its subconsultants because they must redesign a feature of work to accommodate the owner-directed change.

The next cost risk is the potential cost of rework by the construction forces that are working on the given feature. In both cases it is important to institute a method for collecting the losses of both design and construction progress and quantifying them in a manner in which they can become part of the design-builder's change order proposal. It is also important for the design-builder to be able to demonstrate to the owner that it made every effort to mitigate the damages suffered because of the change.

The next type of change is one that results when the design-builder's design is rendered unsatisfactory due to a change in legislation or governing regulation. The scenario here is that both the DB RFP and the winning proposal were predicated upon the rules that were in effect when the contract was awarded. Thus, there is no "fault" to be found by either the owner or the design-builder. However, powers above the owner decided to change the given rule and now the owner must require that the ultimate design and construction conform to the new rule. Often changes of this nature are related to environmental aspects of the project. They may also be related to evolving changes in local building codes. In this case, the change operates in the same manner as an owner-directed change, and the actions taken are identical to the ones described in the previous paragraphs.

One of the most difficult types of changes to reconcile is the third one on the above list, changes due to an adverse code interpretation. These occur when the design-builder's design is found to not comply with some specific design code referenced in the contract. A good example is the requirement for structural systems to be designed in accordance with the locally applicable seismic code. These codes are highly technical and often open to interpretation. It is the design-builder's responsibility to design the project in compliance with all applicable codes and standards. Often certain design assumptions are made during the preparation of the DB proposal, which then drive the contract value. When one of the assumptions is found to be wrong, it usually adversely affects the design-builder's budget and/or schedule. Nevertheless, because DB contracting assigns singular responsibility to the design-builder for both design and construction, an adverse code interpretation is really evidence of a design error or omission, and the owner should bear no risk for this type of change. These issues could be missed during design reviews and be found after construction has started, or even picked up during final inspection. An example would be the omission of the requirement to install seismic anchors for utilities in a large building.

The final type of change is one that occurs because the design-builder has decided to proceed with construction of a feature of work whose design has not yet been released by the owner for construction and whose final design is different in some way from the design that was assumed when construction on the feature began. In this case, there are two options. First, the owner and the design-builder can evaluate the technical or operational impact of changing the design to coincide with the constructed or partially constructed product. This may be possible and, if it is, this option minimizes the impact on redesign or rework at the design-builder's expense. The other option requires the design-builder to alter the constructed product to conform to the final approved design at its own expense. In this case, the procedures will be followed as described in the next section of this chapter.

Contractors can always proceed at their own risk, but they must expect that from time to time the risk will work against them. To ensure that relations with the owner are not damaged, the design-builder should create a chain of communication with the owner and its staff to notify the owner when the contractor has decided to proceed with construction on a feature whose design has not been finalized. At the same time, the design-builder's staff should develop a contingency plan for the construction that minimizes any potential rework if the final design changes. The final angle is that design-builders should only proceed at their own risk when there is a measurable benefit to the project. Often design-builders will include a special contingency to furnish a source of funds to cover this and the other three types of changes discussed above.

DB Change Order Process during Construction after Final Design Is Complete

Differentiating this discussion from the previous one is the idea that here the final design has been completed and released for construction by the owner. The flow chart for the DB change order process during construction is shown in Fig. 7-3. In this case, the major difference between DB and DBB is that the design-builder will accomplish any redesign effort that needs to be done to execute the change. Additionally, it is the design-builder's responsibility to generate alternatives that mitigate both cost and time effects of the required change. Figure 7-3 shows the design-builder iterating through the process of seeking technical alternatives that allow the project to proceed with the minimum impact wherever possible. This often requires that design subconsultants and trade subcontractors be brought into the alternative development process to assist in finding the alternative that has the least impact.

The most common mistake made by owners in this situation is to not ask the design-builder to define the details of the change. This is a throw-back to the DBB culture where the owner conducts the technical evaluation and directs the construction contractor to change the design per its detailed instructions. In DB, by allowing the design-builder to generate technical alternatives wherever possible, the owner is reinforcing the shift of design liability to the design-builder. Issuing a *directive* change order could potentially alter the ultimate design in a manner that transfers the liability for the performance of the changed feature of work from the design-builder to the owner. But allowing the design-builder to generate technical alternatives also allows for the maximum level of creativity to be applied to the change by not restricting the design-builder's prerogatives. In this way, the owner is not putting itself in a position of creating a delay to the project schedule while it analyzes the change and develops the final directions.

On the design-builder's side, the most common mistake is to not thoroughly explore potential alternatives with its subconsultants and subcontractors. Failing to allow the entities that are going to be affected by the change to make input, risks creating unintentional costs and schedule disruptions. It also eliminates potentially innovative solutions to the change order problem that could indeed reduce both time and cost growth in the project. Finally, the design-builder needs to demonstrate to the owner that it has taken every possible avenue to mitigate the impact of the change to the project. This is essential to maintain the level of trust that is

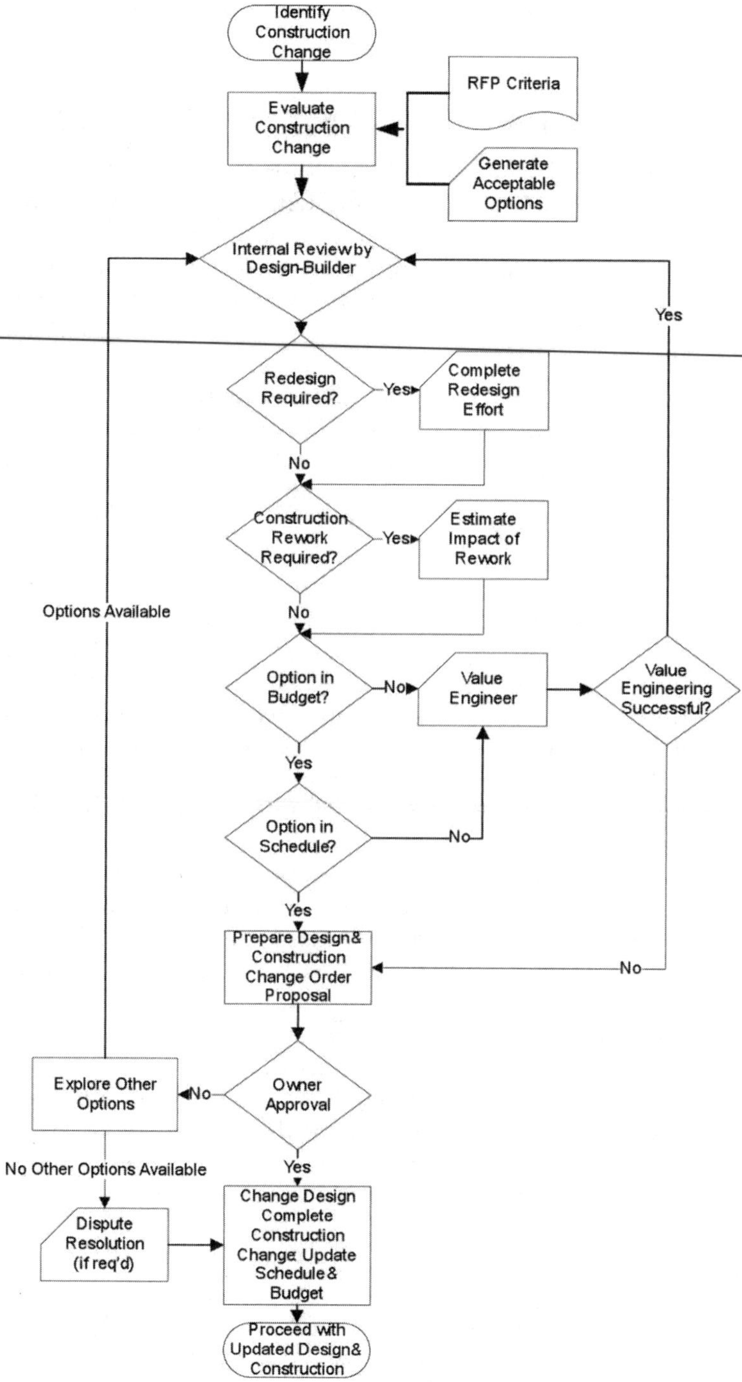

Figure 7-3 DB change flow chart during construction after the final design is complete.

necessary for a successful DB project and to put the design-builder in a position to be competitive for the owner's next DB project.

Changes to Scope, Unforeseen Conditions, Design Errors and Omissions

One of the myths about DB is that it allows the owner to transfer all the risk for changes to the design-builder. This is not true. DB contracts follow the same train of logic regarding changes that any design or construction contract follows. This idea revolves around the quantified scope of work portrayed in the original contract. Thus, anything that increases or reduces that scope of work constitutes a change to the contract. In DB, the owner is asking for both design and construction work to be completed under the contract. Therefore, any change to the amount of design or construction effort required would be classified as a change order. As a result, in determining the requirement for a change order, one must first determine which party to the contract is initiating the change and whether the changed scope affects the contract price or schedule.

Scope Changes

Changes to a project's scope can come from a variety of sources, many of which have already been described in previous sections of this chapter. The issue of a scope change must necessarily cover both the design and construction processes. Changes to the design scope are much more difficult to recognize and, hence, quantify than changes to the construction scope. Design is inherently an iterative process, and the design-builder's design team should have priced a reasonable amount of iterative redesign effort as part of the original contract's price proposal. Therefore, the issue then becomes differentiating between redesign effort that is part of the typical design process (i.e., that work necessitated by the DOR deconflicting the design products of various disciplines), and redesign that was completed to accommodate a technical change to the scope of work. The concept of design commitment, as discussed in our previous book (Gransberg et al. 2006) as well as in Chapter 3 and especially Chapter 8 in this book, is useful in this case.

Design commitment is the point at which the DOR has approved the design of a given feature for construction and the design is ready for the owner's final review and acceptance. Essentially, it means that the design is complete and will be built exactly as described in the documents to which the design-builder is committing. Thus, any redesign required to accommodate a scope change after design commitment has occurred would indeed be compensable.

Unforeseen Conditions

Changes due to unforeseen conditions are always the responsibility of the owner (Beard et al. 2001). There have been a number of efforts to draft contract language to shift that risk to the design-builder; for the most part, all have been unsuccessful. The most common type of risk-shifting clause is one that requires the design-builder to conduct a thorough examination of the site and account for any differences between actuality and the scope described in the design-builder's proposal. While this may seem to be a logical way to assign this risk to the design-builder, it has been found to apply to only those aspects that could actually be found during a site visit. A diligent owner will normally have already accounted for all the visible and discoverable site features in the RFP, so the unforeseen conditions will still remain to be discovered during contract execution. Additionally, because owners often select DB project delivery to compress the delivery period (Songer and Molenaar 1996), it is also possible that the owner will not have had the time normally available to conduct thorough site investigations prior to advertising the contract. This makes changes due to unforeseen conditions a very real risk.

Potential changes due to unforeseen conditions should be on the agenda for the first pre-work meeting. Setting up a system to identify and quantify the impact of unforeseen conditions is the best way to mitigate this risk. The owner could also require the design-builder to conduct detailed site investigations at the earliest possible point in the contract to mitigate the impact of unforeseen conditions by discovering them at a point where minimal design and/or construction effort has been expended. A typical example would be a building renovation project where the owner is unsure of the presence of asbestos. In this case, the owner might direct the design-builder, before expending any other effort, to conduct destructive testing that could not be done when the building was occupied. This would allow any asbestos to be discovered and any remediation efforts to be included in the project at a point where the impact is minimized.

Design Errors and Omissions

Shifting design liability to the design-builder is one of the benefits most cited in the literature for selecting DB project delivery (Beard et al. 2001; Gransberg et al. 2006). The very nature of the contract allows this risk to be transferred with clear responsibility. However, the major issue at stake in this category is determining which party to the contract made the design error or omission. There are three major points in the life cycle of a DB contract where design errors and omissions can be made:

- Design errors and omissions in the DB RFP
- Design errors and omissions in the design-builder's winning proposal
- Design errors and omissions in the completed construction documents

The rule of thumb for changes caused by design errors and omissions is: "Whoever designed is liable for it." Thus, despite contract language developed to shift this risk to the design-builder, the owner is liable for anything that has been "designed" by the owner or its consultants, including errors that were part of the DB RFP. This is because the design-builder had to rely on that document to prepare its technical and price proposal. Therefore, if a design error in the RFP technical scope is found after contract award, it will necessitate a change in the scope of work that was contemplated in the price proposal, making the change compensable to the design-builder. An example of this situation is an RFP where the owner has directed that all improvements be built on existing right-of-way but during the design phase the DOR finds that this is impossible for a given feature of work. Therefore, the owner will be liable for any extra design and construction effort that may be required, as well as any delays that are induced by the error.

The next place where design errors and omissions can occur is in the design-builder's proposal. Often these errors are associated with the design assumptions made to arrive at conceptual quantities of work used in a lump-sum price proposal. Obviously, this is a risk borne entirely by the design-builder. A typical error of this nature is to underestimate the quantity of steel reinforcing used in a concrete feature by using parametric design formulae. For instance, during proposal preparation a structural designer might direct the estimator to use 150 pounds of reinforcing steel per cubic yard of concrete to arrive at the total amount of steel in the feature. After award, the final design required 175 pounds of steel per cubic yard of concrete. The use of an incorrect parametric design formula was in fact a design error, even though the design was at an early stage of development. The design professionals cannot reasonably expect the construction estimators to know how much that number might grow during the normal iterative design process because those estimators are not licensed design professionals. In this case, the liability for the increased quantities of steel would fall on the design professional who furnished the erroneous formula. The issue will be resolved in accordance with the internal agreements of the design-builder's team.

There are two ways for the design-builder to ameliorate the risk of design errors and omissions in the proposal. First, teaming agreements executed prior to proposal preparation should address this issue and spell out how it will be resolved if it occurs. One method is to require the design professionals who furnish input to the proposal to include design allowances for potential quantity growth during the normal design process, based on their experience. In the previous example, the designer might have indicated that the quantity of reinforcing steel to be in the range of 150 to 175 pounds per cubic yard of concrete, with the extra 25 pounds constituting the design allowance. It would then be up to the design-builder to decide whether to use the lower or the higher number in its price proposal. In this method, the designers would then be made aware of the "as-proposed" or "as-bid" quantities of work and be instructed to continually check their work against these quantities to identify when their designs are evolving to a point where they will exceed them. When that point is reached, the design professional must then notify the design-builder and the team can then look at alternatives to mitigate the possible financial loss due to the erroneous proposal design input.

The second method is to create a specific contingency to cover design errors and omissions in the DB proposal. This method assumes that there will be errors and seeks to manage the risk by establishing a fund that can be used to pay for the impact of those errors. Setting the contingency will require the design professionals to furnish the design-builder with a detailed listing of those areas where scope creep is normally observed and the potential impact of those increases. The teaming agreement should also address the details of how this contingency will be allocated during contract execution.

The third area is design errors and omissions that are contained in the design-builder's final construction documents. While intuitively this may seem to be the responsibility of the design-builder, the source of the error must be carefully analyzed before liability is assigned. The major issue is usually the impact of the owner's design review comments. An owner can unintentionally shift the design liability back to itself by the nature of its design review comments. Review comments that direct the design-builder to change the design in some significant manner, causing the applicable feature to fail to perform as expected, are the responsibility of the owner, not the design-builder. This situation requires an owner to train its technical staff to record review comments in an *advisory* rather than a *directive* manner. For instance, rather than directing the designer to increase the size of a footing to comply with stated contract factors of safety, the owner should indicate that the subject footing appears not to satisfy the requirements and cannot be accepted until it does. This puts the onus on the design-builder to determine how much the footing should be increased, preserving the transfer of design liability as intended.

The other situation in which liability for design errors and omissions in the construction documents is not clear is found when the error can be traced back to the design products that were developed by the owner when it prepared the DB RFP. There are two sides to this issue. On one hand, it is arguably the design-builder's responsibility to take the incomplete design expressed in the RFP and advance it to the point where construction documents can be produced. Therefore, it is part of the normal design development process to identify and correct errors or omissions in early designs, which makes those errors and omissions the design-builder's responsibility. The other side of the debate maintains that the design-builder had to rely on the RFP design documents to prepare its proposal and thus arrive at a lump-sum price. Thus, because there was no time during proposal preparation to verify the sufficiency of the owner-provided design, it is the owner's responsibility to compensate the design-builder if a design error or omission is found post-award. Both sides in this battle have valid arguments based on case law, and it is impossible to generalize on this particular issue. Suffice it to say that the easiest resolution is for the owner to carefully review the design products contained in the RFP and ensure that they are indeed correct.

Change Order Costs

Estimating the cost of change orders in DB projects can be quite difficult, depending on the stage of the project at the time the change is identified. While changes that occur during construction usually have the greatest impact, they are far easier to quantify than changes implemented during early design stages. This is because the estimator can use the construction documents as the benchmark against which to measure the effect of the change. Thus, it is imperative that the estimator fully understand the scope and impact of a change to a DB contract when developing the change order cost estimate. Scott Cullen of the National Institute of Building Sciences put it this way:

> . . . in order to properly prepare an estimate for a change order, the estimator must understand the overall objective, i.e. to reach an agreement with the contractor that is in the best interest of the [owner]. The lowest possible price does not always meet this objective nor would a "generous" price if that price offers more payment than necessary to include sufficient incentive. (Cullen 2005)

Change order costs need to be accumulated in specific categories to allow both parties to the negotiations to fully understand the logic behind each estimate and to ensure that a direct comparison can be made between the design-builder's change proposal and the owner's independent estimate. Using this approach allows the negotiation of changes to proceed without potentially damaging the important trust relationship that is so essential to a successful DB project. It also allows the identification of gaps and overlaps in the estimate. Some public agencies have specific formats in which contractors must organize their change proposals. Regardless of the administrative procedures mandated for the change process, change order costs can essentially be grouped into the following generic categories:

- Direct costs: costs associated with the specific feature of work being changed.
- Indirect costs: costs associated with overhead at both the project and home office levels which are normally added to direct costs to calculate total cost.
- Schedule-related costs: costs associated with the change in the project schedule.
- Impact costs: costs associated with unchanged work that is affected by the change.

Direct Costs

Direct costs are best subdivided into the direct costs of design and those related to construction. It is possible in a DB project to have only one phase of the project affected by a change. Thus, while it is appropriate to consider how design changes affect construction costs, it is possible that the solution can have no effect at all. The same rule applies to construction changes. If the owner is able to convey the change in a manner that does not require the design-builder to either lose previous design effort or to redesign the given feature, then there will be no direct design costs.

Given the above discussion, direct costs are self-explanatory. The major shift in DB from DBB change order estimating is to ensure that both the design and construction phases are adequately considered. Direct costs will generally be of a nature that is tied to the actual production of the design and construction products that are associated with the change order.

Indirect Costs

DB contracting does not change the concept of estimating indirect costs. These costs are essentially the costs of doing business that furnish the products covered by the direct costs. What can be tricky in DB is the increased number of members of the design-builder's team whose indirect costs might be affected by a change. Because design and construction can be done in parallel, it is common to find construction trade subcontractors furnishing portions of the design. For instance, a licensed engineer working for the electrical subcontractor may complete the electrical design, eliminating the need for the DOR to hire an electrical subconsultant. Thus, it is possible for a "construction" team member to need to add both direct and indirect costs to a change that appears to be purely design.

The other issue in computing DB indirect costs deals with the requirement by many owners for members of the design-builder's team to co-locate during project execution. Co-location offers many advantages but it is not free. In fact, it can add a third level of overhead beyond the traditional job site and home office indirect costs. It also generally increases the indirect costs above typical levels due to the need to supply redundant facilities and services as well as additional administrative personnel whose function would normally be handled at the home office. The easiest way to approach this issue is to view the indirect costs of co-location as merely an extension of the job site overhead costs as they are in fact project-specific.

Schedule-Related Costs

The schedule-related costs meant here are those that directly spring from the change in the scope of work related to the given change order. Thus, these are usually quite straightforward and will be identified during the change order analysis. They will include such costs as extended overheads if the change occurs on the project's critical path; increased labor rates if the change forces the work into a period where union wage scales are increased; premium pay for labor that must work extended periods to achieve critical milestones; and expediting costs for materials and equipment that may have become critical due to the change.

Impact Costs

Impact costs are related to the unchanged work, and identifying them is not only indirect but also quite abstract. One author describes them as "very broad, intangible, and susceptible to a large variety of situations" (Cullen 2005). By and large, impact costs flow from the acceleration or delay of the unchanged work. Quantifying them can be very difficult. Picking a typically abstract example, it is extremely hard to quantify the cost of a project-wide work slowdown due to a pending change order. It is an established fact that subcontractors and foremen will literally reduce production until they know whether their piece of the project will be changed. Once they find out that it will not, full production resumes but a marginal delay has indeed been recorded. In DB, the situation is even more complex in that impact costs must be identified in both the design and construction process. Thus, the process must ensure that all aspects of the DB project are reviewed on a case-by-case basis to identify potential impact costs.

The federal government uses a procedure in this portion of change order analysis to separate impact costs into two categories: factual and judgmental (Cullen 2005). "Factual impact costs" are those where documentation exists to both establish their existence and help bound their value. An example would an equipment rental agreement that was extended due to the change's impact. According to Cullen (2005), typical factual impact costs are:

- Escalation of material prices
- Escalation of labor wage rates
- Change in equipment rates
- Increase from extending the storage period for materials and equipment
- Increase from extending the contract for labor cost and subsistence
- Increase from a longer period of equipment rentals or use
- Increase from a longer period of utilizing overhead personnel, materials, and utilities
- Increase from a longer period of providing overhead and project office services

To this should be added:

- Travel costs to bring specialty subconsultants to the project site to evaluate design alternatives for the change in light of existing field conditions

"Judgmental costs" are defined as "those that are dependent on variable factors such as performance, efficiency, or methodology and cannot be stated factually prior to actual accomplishment. These must be negotiated and based upon experienced judgments" (Cullen 2005). Negotiating these costs will greatly depend on the credibility of the DB team and the strength of the mutual trust at the time of the change order. A previous study of partnering in DBB projects underscored the value of executing the project in a trustful business relationship (Gransberg et al. 1999). That study found that partnered projects had more change orders than those that were not partnered, but that the average cost of a partnered project's change order was less than half of that found on a non-partnered project. That led the authors to conclude that the owner's willingness to consider contractor-requested changes was returned by the contactor keeping the costs as low as possible. The lesson for DB change order negotiation is for the owner to ensure that it is willing to consider any and all judgmental impact costs during negotiations, and not reject these types of impact costs out of hand because no specific documentation can be produced to factually support their existence.

The paper by Cullen (2005) does an excellent job of describing the process of identifying and calculating judgmental impact costs, and lists typical ones as:

- Change of efficiency resulting from rescheduling
- Loss of labor efficiency resulting from longer work hours
- Loss of efficiency caused by disruption of the orderly existing processes and procedures
- Inefficiency from tearing out completed work and the associated lowering of morale
- Loss of efficiency during rescheduling of manpower
- Inefficiency incurred from resubmittal of shop drawings, sample materials, etc.
- Additional costs resulting from inability to transfer manpower expertise to other work (Cullen 2005)

To this should be added:

- Increased design cost due to the need to use senior designers due to time constraints.

The bottom line for DB change order costs relies on the careful and fair analysis of all the components of the change in light of the level of design completion at the time the change is identified. The need for a strong level of trust in contract execution is most keenly felt when there is a need to change the scope of work. At these times the design-builder must work hard to ensure that the cost, time, and impact of the change are mitigated, and to act as an advocate of the owner as it defines the actual technical and business aspects of the change. On the other side, the owner must control its level of anxiety and come to the table with an open mind. It should adopt an attitude that it is willing to pay all allowable and allocable costs that the design-builder is able to support through either factual or judgmental means.

Value Engineering Changes

Chapter 3 cited the following Federal Acquisition Regulation (FAR) definition of VE, but it is germane to repeat it here:

> The systematic application of recognized techniques by a multi-disciplined team to identify the function of a product or service, establish a worth for that function, generate alternatives through the use of creative thinking, and provide the needed functions to accomplish the original purpose of the project, reliably, and at the lowest life-cycle cost without sacrificing safety, necessary quality, and environmental attributes of the project. (Blanding 2005, p. 3)

VE is required on all federally funded projects, and the government requires that a VE study be conducted during one of the following phases of project development:

- Initial concepts
- Comparison of alternatives
- Final design (Blanding 2005)

Of the above three phases, final design is the one that is certainly included in the execution period of a DB contract. The initial concepts and comparison of alternatives phases typically occur before DB contract award. However, it is possible that alternatives may be generated by the design-builder post-award, specifically for VE. Therefore, the typical progression of a design-build project encourages the long-standing VE principle that savings are maximized by conducting VE as early as possible. A white paper produced by The Haskell Company states, "Value engineering works best as a collaborative, integrated process, not a set of singular examinations" (The Haskell Company 2008). In DB project execution, the earliest point would be at the pre-work conference conducted shortly after contract award. Our previous book recommends that an agenda item for this first conference be the consideration of alternatives not contemplated in the RFP or the winning proposal (Gransberg et al. 2006).

When dealing with VE change order costs, one must have a baseline against which the change in cost, time, and/or impact is measured. There is some controversy regarding the ability to actually conduct VE in a DB project. Two researchers explain the issue like this: "Design-build contracts do not usually include value engineering clauses for the simple reason that the owner has not provided a design that can be value engineered" (Cushman and Loulakis 2001). This school of thought maintains that all the VE that can be done in DB should be accomplished before the RFP is released, and that any remaining VE will be done by the design-builders in their competitive proposals. This ignores the possibility of the owner unintentionally—through ignorance or omission—eliminated a possible alternative that would in fact add project value. It also erroneously defines the "design" as only the completed construction documents and exposes a DBB mentality that requires a completed design that benchmarks any VE savings.

Fortunately, the literature contains a number of examples of successful VE during DB contract execution. According to Lewis & Zimmerman Associates, Inc. (2008), VE in DB should be conducted throughout the DB project's life cycle by the party that controls the details of design at any given stage. In this approach, the owner should consider conducting VE studies during the following stages:

- Concept design
- Completion of preliminary design
- RFP documents before advertising
- Submitted proposals before award

The design-builder then conducts VE studies post-award at the following stages:

- VE of the proposal demonstrating specific action to improve value on the project
- VE of the project to develop Value Engineering Change Proposal opportunities, as permitted by the Owner (Lewis & Zimmerman Associates, Inc. 2008)

Given the above discussion, VE change order costs would be accumulated in much the same manner as normal change orders, using the most current design documents as the baseline against which to compute both costs and savings. Most federal DB contracts include a VE clause to allow the government to capture potential savings throughout the DB project's life cycle, and do not limit VE to the design phase.

To summarize this discussion of change management, change is inevitable and thus both owners and design-builders must enter the DB project armed with a plan to manage change as it occurs. Successfully implementing a change management plan requires that the relationship between the two parties be such that precise, accurate communications of actual project conditions and events can be transmitted without fear of reprisal. Research has shown that the partnering process helps develop this level of trust, and assuming an "open books" attitude toward sharing the cost, time, and impact data that make up the body of a change order proposal helps reinforce the credibility of both parties. DB is not a magic bullet that automatically shifts the liability for changes to the design-builder. However, the qualifications-based, best-value selection of the design-builder creates an incentive for the design-builder to accommodate the owner's preferences and mitigate the impacts of unavoidable changes as a means of enhancing its competiveness for the owner's next DB project.

References

Allen, L. N. (2001). "Comparison of design-build to design-bid-build as a project delivery method." Master's thesis, Naval Postgraduate School, Monterey, Calif.

Barrie, D. S., and Paulson, B. C. (1991). *Professional construction management*, McGraw Hill, New York.

Beard, J. L., Loulakis, M. C., and Wundram, E. C. (2001). *Design build planning through development.* McGraw-Hill, New York.

Bennett, J., Pothecary, E., and Robinson, G. (1996). *Report, design and building a world class industry: Reading Design-Build Forum*, Center for Strategic Studies in Construction, University of Reading, Reading, UK.

Blanding, J. R. (2005). "Using VE in design build." *Proc., 2005 AASHTO Value Engineering Conference, San Antonio, Texas*, American Association of State Highway and Transportation Officials, Washington, D.C.

Construction Industry Institute (CII). (1994). "Project change management research team." *Special Publication 43-1, US.* The University of Texas, Austin, Tex.

Cullen, S. (2005). "Construction phase cost management." *Whole Building Design Guide*, <http://www.wbdg.org/resources/construct_cost.php> [July 1, 2008].

Cushman, R. F., and Loulakis, M. C. (2001). *Design-build contracting handbook,* Wolters Kluwer Law & Business/Aspen Publishing, New York.

Federal Highway Administration (FHWA). (2006). "Design-build effectiveness study: Final Report to Congress as Required by TEA-21." *USDOT Federal Highway Administration,* <http://www.fhwa.dot.gov/reports/designbuild/designbuild0.htm> [Aug. 30, 2006].

Friedlander, M. C. (2007). "Legal issues unique to design-build." Schiff Hardin, <http://www.schiffhardin.com/binary/legal_issues_design_build.pdf> [September 15, 2007].

Gransberg, D. D., Koch, J. E., and Molenaar, K. R. (2006). *Preparing for design-build projects: A primer for owners, engineers, and contractors,* ASCE Press, Reston, Va.

Gransberg, D. D., Badillo-Kwiatkowski, G. M., and Molenaar, K. R. (2003). "Project delivery comparison using performance metrics." *2003 Transactions,* AACE International, Orlando, Florida, DRM 02.01–02.06.

Gransberg, D. D., Dillon, W. D. , Reynolds, H. L., and J. Boyd, J. (1999). "Quantitative analysis of partnered project performance." *ASCE J. Const. Eng. Mgmt.* 125(3), 161–166.

Hale, D. R., Shrestha, P. P., and Gibson, G. E. (2009). "Empirical comparison of design/build and design/bid/build project delivery methods." *ASCE J. Const. Eng. Mgmt.,* 135(7), 579–588.

Hallock, B. (2006) "Managing change vs. administering the change order process." *Nielsen-Wurster Communique,* The Nielsen-Wurster Group, Inc., 1(6), 1-9.

Haskell Company. (2008). "Value engineering in design-build." The Haskell Company, <http://www.thehaskellco.com/upload/NewsLibrary/WhitePapers/ValueEngineering.pdf> [July 3, 2008].

Konchar, M., and Sanvido, V. (1998). "Comparison of U.S. project delivery systems." *ASCE J. Const. Eng. Mgmt.,* 124(6), 435-444.

Lewis & Zimmerman Associates, Inc. (2008). "Value engineering in design/build." *Lewis & Zimmerman Associates, Inc.,* <http://www.lza.com/lza/vs/design_build.html> [July 3, 2008].

Reed, C. B. (2003). "KPMG construction cost review California State University, Dominguez Hills – 431P Welch Hall – Information Technology Building." *California State University,* <www.calstate.edu/audit/Audit_Reports/construction/2002-2003/DominguezHillsConstruction.pdf > [March 12, 2010].

Riley, D. R., Diller, B. E., and Kerr, D. (2005). "Effects of delivery systems on change order size and frequency in mechanical construction." *J. Constr. Engrg. Mgmt.,* 131(9), 953–962.

Runde, D. F., and Sunayama, Y. (1999). "Innovative contractor selections methods: Alternatives to the traditional low bid in Massachusetts public construction." *Policy Analysis Report,* John F. Kennedy School of Government, Harvard University, Boston, Mass.

Songer, A. D., and Molenaar, K. R. (1996). "Selecting design-build: Private and public sector owner attitudes." *ASCE J. Eng. Mgmt.,* 12(6), 47–53.

Warne, T. R. (2005). "Design-build contracting for highway projects: A performance assessment." *Report to the California Design-Build Coalition,* Tom Warne & Associates, LLC, South Jordan, Utah.

Zou, Y., and Lee, S. H. (2006). "Impacts of project change management on project cost and schedule performances." *2006 AACE Int. Trans.,* AACE International, Morgantown, W.V.

EIGHT

Progress Payments during Design and Construction

Every action that takes place in executing a DB contract revolves around the design-builder satisfactorily completing some requirement so that it can get paid, or the owner determining that a given requirement has satisfactorily been completed so that it can pay the design-builder. The design-builder's proposal will be predicated on an expected cash flow of progress payments; if that is not realized, the design-builder will begin to incur additional costs of financing that will erode the project's profit margin. One author describes the impact to a DB project when actual cash flow does not mirror planned cash flow:

> Too often on large projects, design-builders get behind in their compensation from the owner because of unresolved differences of opinion. . . . When this happens, parties can quickly become adversarial instead of working together for the good of the project. With financial difficulties caused by cash flow problems, genuine problems with the project can develop as a result. (Holland 2002)

Therefore, understanding the payment process and how it affects the DB contract administration is critical to the success of the project.

First, one must understand the motivation for progress payments in all types of project delivery methods. Owners make progress payments to avoid paying the cost of design-builders having to finance the entire project themselves. In the public sector, owner agencies are often not allowed to award a DB contract unless the funding for the project is authorized and available. Thus, because the funds for the project are on hand, it makes sense to leverage them to reduce overall project cost by making them available to the design-builder as satisfactory progress is made. These regulatory restrictions are not normally in effect in the private sector, but private sector owner/developers do not enter into major capital investments such as facility construction unless they are assured of financing. Again, using these funds to make progress payments will reduce project carrying costs (i.e., the financing cost to the design-builder after DB contract award).

Progress payments are normally made after the satisfactory completion of contract design and construction work packages. As such, they constitute the DB project's financial progress and must necessarily reflect the project's physical progress. As a result, they constitute a powerful project control tool by converting the progress of all of the project's work into a single common unit of measure: the dollar. Figure 8-1 shows how the concept of progress payments relates to the DB project quality assurance (PQA) universe discussed in Chapter 4. Essentially, progress payments are made when the owner is willing to allow design products to be released for construction, and when constructed products are inspected and found to comply with the approved design.

The major issue from the owner's perspective in determining the amounts of progress payments is to protect itself against the potential for design-builder default. The owner must ensure that it does not overpay the design-builder at any point in the project because, if default occurs, the project will not have sufficient funding to be completed. On the other hand, if the owner is too conservative in its published payment scheme (e.g., mandating a large percentage of retained earnings), it will force the design-builder to finance more of the project's cost and the project's proposed price will thus be increased to include the carrying costs. Therefore, the owner is treading in a narrow range defined by the maximum amount it should pay

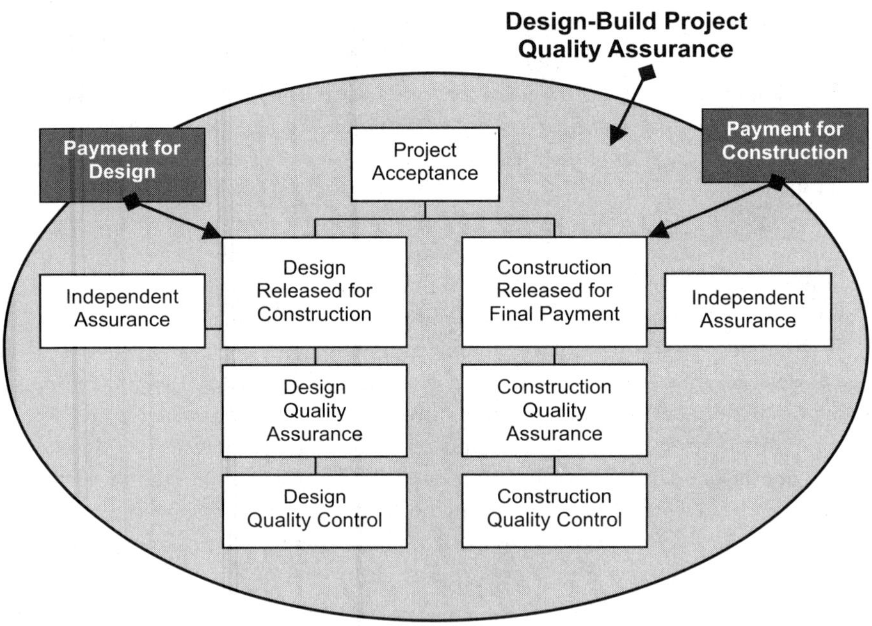

Figure 8-1 DB project quality assurance universe.
Source: Adapted from Gransberg et al. (2008), Figure 3.

at any point in time to manage default risk, and the amount it can afford to pay for financing charges that do not put bricks and mortar into the project. The good news in DB contract administration is that the use of a best-value award method reduces the risk of default through the prequalification process used to select a design-builder. The owner gets to make a qualifications-based selection of both the designer and the constructor. Thus, the potential that an incompetent or unqualified constructor will win the job is significantly decreased. This allows the owner to be less conservative when designing its DB progress payment scheme, which should be reflected in marginally lower project costs. The major mistake made by many public owners is to roll their DBB contract payment clauses over into their DB contracts, thus unwittingly losing the cost benefit of the best-value award.

Most DB projects utilize the "Schedule of Values" as the contractual basis for calculating progress payments. According to Sidney G. Saltz, a senior counsel with Holland & Knight, LLP, "A Schedule of Values is a compilation of the values of the various phases of the work, so that requests for payment to the design-builder can be evaluated" (Saltz 2007, pp. 8-11). It essentially requires the design-builder to assign a value or budget for every activity on the DB project's schedule. Then, upon owner review and approval, it allows the development of a periodic pay estimate to be made for those activities that were underway during the pay period. Simply put, the design-builder can ask for 50% of a given activity's value if it can prove to the owner that the activity is half completed and the work done so far complies with the contract quality requirements. Owners, designers, and builders in the architectural construction area have long been familiar with the Schedule of Values and how to implement it effectively. However, those in the heavy civil arena are accustomed to unit price contracts, where measured quantities of unit price pay items are used to compute progress payments and where the risk for quantity overrun is held by the owner. Therefore, the transition to a lump-sum DB contract and the Schedule of Values must be carefully managed to ensure that all parties understand the financial implications to their standard business practices.

This chapter will cover the six major components of the DB payment scheme:

1. Payment for design progress.
2. Payment for construction progress.
3. Retainage and final payment.
4. Limitation on payments.
5. Incentive payments.
6. DB project cost control.

Payment for Design Progress

Paying for design progress is fundamentally different from paying for construction progress, in that intellectual property is developed in design whereas tangible property is developed during construction. Complicating the issue is the fact that design is inherently an iterative process where the designer-of-record (DOR) must reconcile the design products of each discipline with the design products of all the other disciplines. This usually necessitates a certain amount of redesign by most of the design professionals on the DB team. This is part of the normal process, and its cost should have been included in the value of the design services included in the DB price proposal. This begs the question of how an owner can know when the design of a given element is satisfactorily completed if it might be changed later in the design process to deconflict it with another discipline's output. On the design-builder's side, the issue is one of generating sufficient cash flow during the design process to support the efforts of the design team while ensuring the integrity of the design process. In our previous book, the concept of "design commitment" was proposed as a solution to this conundrum (Gransberg et al. 2006). It is also discussed in Chapters 3 and 7 in this book as well as in the next section.

Design Commitment

Design commitment is a concept that permits the owner to manage the design responsibility during project execution. As previously stated, the design-builder effectively "owns" the details of design during project execution. Thus, it can change working drawings to accommodate budget and schedule constraints. This is particularly necessary in DB projects that are formed using a firm, fixed price or lump-sum contract because the owner is forcing the design-builder to commit to a price before the details of design are known. However, to make progress payments for satisfactorily completed construction, the owner must have a finished design document to be able to determine the physical percentage of completion. In other words, How can the owner determine that the concrete in the foundation is 20% complete when the design of the foundation is subject to possible change? Therefore, the concept of design commitment has evolved to facilitate the progress payment process.

Design commitment is made when a working drawing and/or specification becomes a construction document approved by the DOR for construction. In other words, the DOR is indicating that no further design work is contemplated and that this feature of work is ready for final owner review. Additionally, the design-builder is indicating that the committed design product has been priced and that the work can proceed according to the approved schedule. Thus, the owner's representative has a yardstick with which to measure the construction progress and authorize progress payments. Any change after design commitment requires a bilateral agreement. Before design commitment, either the DOR or the design-builder can make changes as required to meet technical, budget, and schedule constraints as long as they comply with the technical requirements contained in the contract. Thus, design commitment essentially acts as a payment mechanism as well as a means to communicate the progress of design effort in a DB project.

Payment Formula for Design Completion

In order to make a payment for completed design, the DB contract must have a formula to pay for design progress and a method to measure its timeliness. The owner could require that the design-builder break out design fees in the Schedule of Values, either by major feature of work (e.g., foundation, superstructure) or by milestone (e.g., 50% design submittal, 90% design submittal, and so forth for an engineered project; or conceptual design, schematic design, and so forth for an architectural project). A third method is to pay by the completed sheet of design drawings. The first method rolls the design activities nicely into the construction schedule and makes paying for construction on the condition that design is complete and visible on the project schedule itself. The second method seems to be the more commonly used because it is used by both owners and designers in a DBB design contract. The third method turns the design process into a unit price environment where the bias is toward completing as many sheets as possible to maximize the design fee; this is not recommended.

Regardless of the design package designation method, the design payment clause should require that design commitment for a given feature of work be required to claim the full design fee allocated to that feature. This can be easily demonstrated by construction drawings and specifications being complete and approved for construction by the DOR. This approach is carried over into the payment for construction. The clause should state that no payment for construction would be made unless the design is complete and design commitment for the given feature of work has been made. Additionally, all requisite shop drawings and submittals must also be approved by the DOR. Box 8-1 also shows how one federal agency worded those clauses to fit with its design commitment clause.

Determining an accurate activity completion percentage for a design activity is complex, if not impossible. The lack of tangible benchmarks puts this squarely into the realm of professional judgment. However, the owner can use the output of the design QA program discussed in Chapter 4 to assist in determining design activity progress. For example, a design submittal that has been reviewed and has no significant deficiencies to correct indicates that the activities necessary to complete it are nearly finished. The opposite is true for a submittal in which owner review revealed a substantial need for design work to make it compliant with contract requirements. This will be easier to do if the design submittals consist of discipline-specific work packages rather than design milestones. To gauge the level of design completion, the owner can ask the following questions:

- What specific design information does the builder need to have to properly construct?
- How long will it take to design each feature?
- At what point during the design process can construction of each feature begin?
- How much design will be done by trade subcontractors in submittals?

A design submittal that contains enough design detail to permit it to be constructed and perform as required by the contract is nearly complete. The length of time required to complete a given design submittal is proportionate to the number of disciplines that must contribute to its content and be coordinated. Thus, a

Box 8-1 Sample Roles and Responsibilities of the Designer-of-Record, and Design Commitment and Design-Build Payment Clauses

H.8 RESPONSIBILITY OF THE CONTRACTOR FOR DESIGN

(a) The Contractor shall be responsible for the professional quality, technical accuracy, and the coordination of all designs, drawings, specifications, and any other non-construction services furnished by the Contractor under this contract. The Contractor shall, without additional compensation, correct or revise any errors or deficiency in its designs, drawings, specifications, and other non-construction services.

(b) Neither the Government's review, approval or acceptance of, nor payment for, the services required under this contract shall be construed to operate as a waiver of any rights under this contract or of any cause of action arising out of the performance of this contract, and the Contractor shall be and remain liable to the Government in accordance with applicable law for all damages to the Government caused by the Contractor's negligent performance of any of the services described in paragraph (a) furnished under this contract.

(c) The rights and remedies of the Government provided for under this contract are in addition to any other rights and remedies provided by law.

(d) Roles and Responsibilities of the Designer-of-Record: The Designer-of-Record (DOR) is the single point of responsibility for all design decisions and design products for the Design-Build Contractor and shall supply the required professional liability insurance. The DOR shall review, coordinate, deconflict and approve for construction all design and extensions of design produced all members of the DB contractor produces it and internal contractual arrangements between members of the DB contractor's team including design subconsultants, construction subcontractors, material suppliers and other entities as required. The DOR shall indicate review and approval on all record drawings, specifications, and . . . by fixing a stamp indicating approval for construction or the DOR's seal as appropriate. The DOR shall ensure that all design product has been priced and found to conform to the requirements of the construction schedule and the DB contractor's budget prior to submitting them to the government for review and before finally approving elements of design work for construction in accordance with the clause entitled "Design Commitment."

(e) The DOR shall conduct and document regular jobsite quality assurance inspections and verify that the contractor's quality control system and construction quality conforms to the record drawings and specifications. The DOR shall verify in writing that all partially completed design and construction is in good order before partial payment is claimed by the DB contractor for those items of work. The DOR shall conduct a prefinal inspection, prepare a punchlist, and then conduct a subsequent inspection to ensure that all items on the punchlist have been corrected PRIOR to the DB contractor scheduling the owner's pre-final inspection.

(Based on FAR Clause 52.236-0023, "Responsibility of the Architect-Engineer Contractor (APR 1984)")

Box 8-1 Sample Roles and Responsibilities of the Designer-of-Record, and Design Commitment and Design-Build Payment Clauses(continued)

H.8a DESIGN COMMITMENT

(a) The DB contractor owns details of design during the project and may change them on working drawings and specifications as required to meet budget and schedule constraints. DB contractor can change working drawings to accommodate budget and schedule constraints.

(b) The DOR shall establish a drawing and specification numbering system that clearly labels each design product and indicates whether the specific item is a working document subject to unilateral change or a final document. Final documents have been priced, scheduled, and approved by the DOR for construction.

(c) Once a document has been identified as a final document, then design commitment will have occurred by the DB contractor. Design commitment means that the feature of work conforms to the DB contractor's budget, schedule, the quality requirements of the contract, and will be constructed as detailed in the final document. Design commitment must be made prior to submitting a feature of work for final review and approval by the government. The DOR will sign and seal all final drawings and specifications.

(d) Any changes to documents for which design commitment has been made will require approval of the government.

H.8b PAYMENT FOR DESIGN

(a) Payment for satisfactorily completed design work will be as determined in the schedule of values associated with the Design-Build Contractor's approved project progress schedule for progress.The Contractor may break out design fees in the price proposal by major feature of work or by design milestone as required by the requirements for the price proposal as detailed in the RFP.

(b) No partial payment will be made until working drawings and specifications for the identified feature of work or design milestone are reviewed and approved for payment by DOR.

(c) The government reserves the right to reduce the design fee claimed on any progress any request if significant changes have been made to working drawings previously claimed for payment.

(d) Design commitment must be made on each identified feature of work to claim the full design fee allocated to each feature.

H.8c PAYMENT FOR CONSTRUCTION

(a) No partial payment for completed construction will be made until the design is complete and design commitment has been made for that feature of work.No payment for completed construction will be made until all requisite shop drawings and submittals have been reviewed and approved for construction by Designer of Record.

Source: USAID (2001).

short duration on a design activity can signify that it is relatively simple and easy to finish, whereas a long duration indicates more complexity and it will not progress as speedily due to the need to coordinate the input of several designers. Next, some features of work, such as electrical and mechanical systems, cannot be released for construction until their design is 100% complete. Others, such as site work, are flexible or simple enough that corrections to comply with the details of their final designs are easy and inexpensive to make, which allows the design-builder to begin physical work before the full design is formally released for construction. Finally, trade subcontractor submittals are design extensions of the construction documents. If a feature of work is essentially going to be defined by the trade subcontractor, it is in the project's best interest to initiate the subcontractor's progress and allow it to complete those critical design submittals. Thus, the "design" for that feature of work will consist of performance specifications and sufficient drawings to permit the subcontractor to develop the necessary detailed design products.

Payment for Construction Progress

If the DB contract is structured to not allow construction to begin before the final design is approved, then payment for construction is no different than it would be in a DBB project. However, because the major benefit of using DB is the ability to compress the delivery period to as short as possible, most DB projects will have design and construction occurring in parallel. This necessitates a different approach to paying for construction progress. This approach requires the following four events to be complete before a design-builder can claim the full value of a given construction activity:

1. The design associated with the construction activity is complete.
2. The design associated with the construction activity has been reviewed and found to be in compliance with the contract.
3. The design-builder has made design commitment for that activity.
4. The constructed product is complete and has been inspected in accordance with the approved QA/QC plan and found to be compliant with the contract.

Determining Activity Progress

By verifying the design commitment, the payment formula for design completion, and the payment formula for construction progress, the owner is ensuring that it can measure 100% completion for that particular activity. By applying this approach across an entire feature of work, the owner has protected itself against unintentionally overpaying for perceived progress, and has put the responsibility for defining the details of construction progress on the DOR where it belongs. Thus, it makes sense to require the DOR to review and validate each periodic pay request to ensure that the construction progress tracks with actual design progress.

If the construction progress gets ahead of the design progress, then the design-builder is proceeding at its own risk and the owner is not obliged to pay anything beyond the work associated with the current approved design. However, this is not to say that the owner does not have the option to pay for the accelerated work if it believes it is satisfactory and will not change as the design evolves.

Obviously, those activities whose design is complete and that are physically finished can be judged to be 100% for purposes of the periodic pay estimate. However, the owner must ensure that all submittals prepared by the trade subcontractors involved in the construction activity have been reviewed and approved by the DOR to make a final determination. The possibility exists in a fast-moving project that the construction will get ahead of the paperwork. Because submittals such as shop drawings and catalog cuts are the final extensions of design detail, it is important to ensure that they are in line with the construction documents. As the DOR produces the construction documents, it is the entity best suited to properly execute this review. Thus it is in the owner's best interest to ensure that this critical step has been taken before agreeing to pay the full value of a construction activity. Additionally, this approach keeps the owner from inadvertently assuming design liability by approving a submittal that constructively changes the DOR's design.

Finally, payment for partial construction activity completion must be inextricably linked to the project's construction QA/QC program. This is the test for adequacy of materials and workmanship. Thus, as is detailed in Chapter 6, the owner may choose to withhold payment on those activities whose quality is in question. Figure 8-2 illustrates the owner's decision process as it reviews the design-builder's periodic pay estimate to determine the appropriate percentage of completion for each construction activity. It shows that there are three possible decisions for each activity. If the activity passes all the gates described above, the owner should authorize payment of the full amount requested for that activity. If design commitment has not been made, the owner can choose to pay a portion of the requested amount if it feels that construction quality is adequate and the feature of work will probably not change during the remaining design process. However, if either of these conditions is not met, then the owner has no assurance that what has been built will ultimately comply with the requirements of the final design. Therefore, it can reasonably reject the request for partial payment on the construction activity. The final issue is the same as in DBB construction administration: the resulting construction product must not have any unresolved quality issues before it is declared complete for pay purposes.

The New Brunswick, Canada, DOT uses a device that links payment for DB project construction project directly to what it terms its Quality Management System (QMS). A paper presented at the 55th Annual Quality Congress details the process as follows:

> A novel feature of the DDB [Developer DB or PPP] Agreement is the identification of Quality Management as a "line item" in the Guaranteed Maximum Price (GMP) of the project. This item, approximately 4% of the GMP, is scheduled as a series of maximum monthly amounts (varying with season

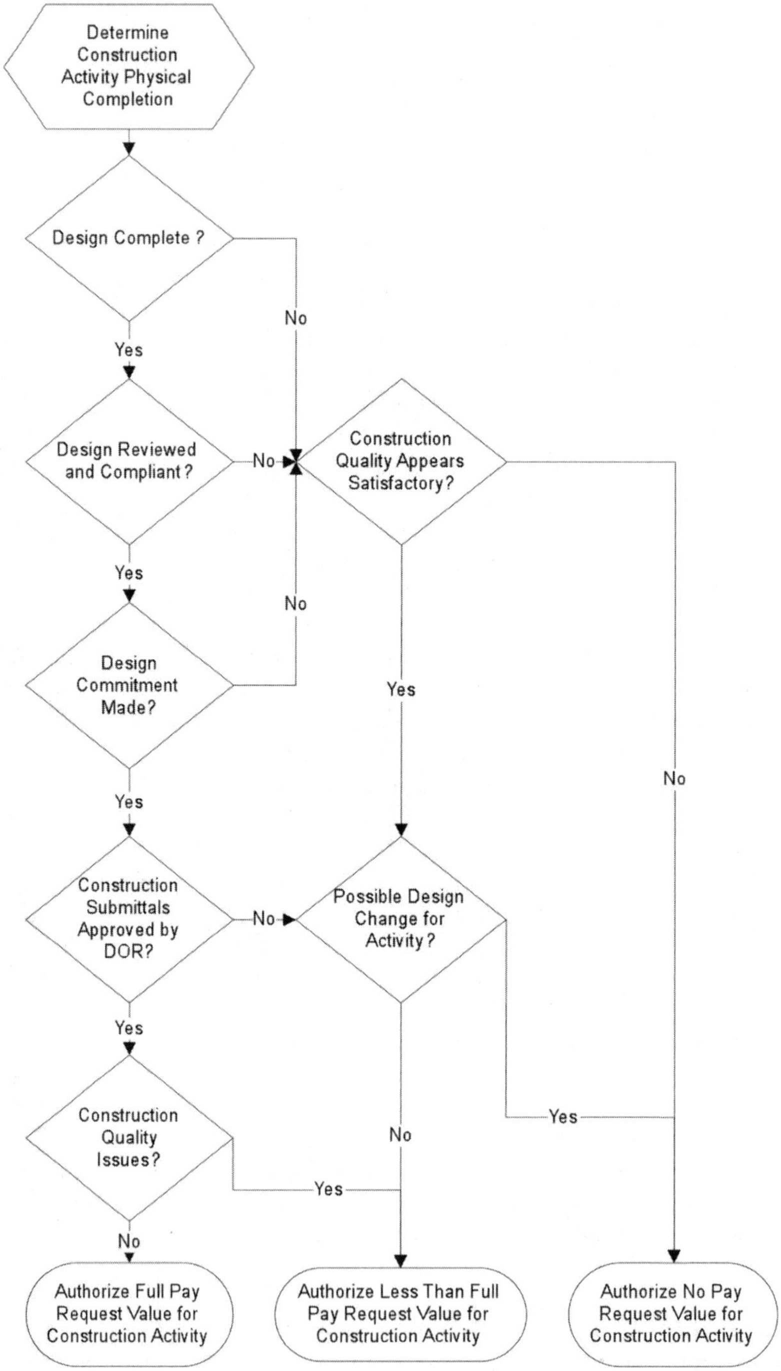

Figure 8-2 Construction activity percentage completion determination flow chart.

> and certain project milestones) to be paid out based on an independent assessment of the effectiveness of the QMS. . . . The method of assessment introduced by the IA [independent auditor] at the commencement of the agreement used a random auditing of an exhaustive list of requirements in the DDB [PPP] agreement. Random samples proportional to the budgeted or estimated cost of fulfilling each requirement were audited each month. The degree of conformance was reported as the total conformances divided by the total month's sample size (% of conforming observations). (Collier et al. 2001)

That Canadian public agency then uses the degree of conformance to determine the percentage of the monthly QMS payment earned. Thus, the developer has essentially 4% of its project cash flow at risk and the aggressive payment algorithm creates a financial incentive to ensure satisfactory quality. The authors of this report go on to conclude:

> To compete . . . in today's environment, a company must be willing to agree to world-class requirements for quality. By aligning quality requirements . . . and incorporating them directly into the agreements, the advantage went to the proponent who [proposal that] would accept extra responsibility and seek to delegate it effectively. This arrangement also allies the Owner and the Developer closely in pursuit of their common interest—the Public–Private Partnership. (Collier et al. 2001)

Payment for Materials

The long-standing practice of allowing payment for materials on-site that have not yet been incorporated into the project must work a bit differently in DB. The owner should ensure that the DOR has inspected and approved the materials as being in compliance with the construction documents as well as any applicable subcontractor submittals. This can be done by requiring a certificate from the DOR to accompany any pay application that requests partial payment for materials on-site. If the routine pay request procedure already requires the DOR to review and validate the request, as suggested above, the certificate for materials can be dispensed with. The purpose for involving the DOR in this transaction is fourfold. First, it ensures that the materials are indeed the ones that the DOR approved for construction. Second, it protects the owner in the unlikely event of default from having paid for materials that do not meet the approved design. Next, it empowers the DOR on the DB team by giving it the authority to influence the design-builder's cash flow. This creates an incentive to maintain good internal relations with regard to the integrity of the design during the construction process. Finally, it forces the DOR to stay involved in the project from start to finish and stimulates it to include an adequate fee for construction administration in its design fee proposal.

In DB projects that involve a significant amount of expensive capital equipment, there is another technique than can be used to reduce carrying costs during construction. In many cases, the design-builder will have to pay up to 50% of the cost of a major piece of equipment up-front when it places the order. If it takes several months before the equipment is manufactured and shipped to the project, the design-builder will have to finance that amount for that period because it has no ability under the typical DB payments clause to request a partial payment until the equipment has arrived on the project site. In an industrial process project such as a water treatment plant, the value of the capital equipment can be a large percentage of the total project cost, creating a significant increase in project cost due to short-term financing. In this case, the owner can require that the order be placed in the owner's name and then pay the design-builder when it presents a certified receipt for the deposit showing that the equipment will become the owner's property. In that way, if the design-builder were to default before the equipment is delivered, the owner will still receive it and can then have it installed by the new contractor.

Retainage and Final Payment

The practice of subtracting a retained percentage of earnings from every progress payment is a holdover from DBB contracting practices. Many public agencies are required by law or regulation to include retainage clauses in all construction contracts. The definition and motivation for retainage is described by one public agency as follows:

> Retainage is a common construction contracting practice whereby a certain percentage of compensation is withheld by the project owner from the general contractor and, in turn, by the general contractor from subcontractors until the project is completed satisfactorily. Retainage is also used as leverage to assure timely completion. (OPPAGA 2000, p. 1)

Most owners to do not realize the impact of retainage on the construction project's bottom line. One study found that the practice of automatically imposing retainage without regard to project progress caused general contractors to raise prices by 2.2%, and subcontractors to increase prices by 3.6% (Bausman 2005). The trickle-down effect on subcontractors multiplies as the number of tiers of subcontractors increases due to "pay when paid" clauses increasing the delay in payment at each tier. In 2005 Jerry Yakowenko of the Federal Highway Administration conducted a survey of public transportation agencies on retainage policies in DB contracts and found no clear trend as to an appropriate amount of retainage on DB projects. Instead, he found that retainage was more a function of "the agency's policies, or state law, than on any particular concerns relating to design-build" (Yakowenko 2005). The same study found that two agencies, the Ohio DOT and Greenville County, North Carolina, did not withhold retainage on DB projects.

Design-Builder's Perspective on Retainage and Final Payment

Before making the decision to withhold retainage, the owner should consider the perspective of the design-builder and its team of designers and constructors. It should also evaluate the possible impact on project financing that will find its way into the DB price proposal. Project risk is manifested in the price of a DB project in a number of ways. First, design-builders will include contingencies for specific risks such as design errors or scope creep. Next, they will delegate the exposure to certain risks through their subcontracting plan, handing off the risk to their subs who in turn reflect those risks in their prices. Design-builders can also account for some risks by purchasing insurance to guard against specific losses. Finally, the project's overall risk is addressed in the profit margin added to the bottom line. That profit margin is indicative of how much money the design-builder feels it needs to make to expose itself to the risk of project failure. Thus, the profit margin is proportionate to the risk.

One risk that must be accounted for is that final payment and release of retainage will not be made in a timely fashion. In DBB projects, it is common for release of retainage to take months after the project's substantial completion while the owner and the contractor haggle over punchlist corrections or when there is an outstanding claim. When one considers the fact that "[t]ypically engineering might carry a profit margin of 15% . . . and construction management and/or construction 4% to 5%" (Weaver and Weston 2001, p. 269), retaining 10% of each progress payment essentially puts the project's entire profit at risk. Assuming the 15% design fee to be about 5% of the total contract would yield a weighted average profit for the entire DB project of 5.5%. Thus, a 10% retainage also requires the design-builder to finance 4.5% of the project's costs as well. Financing rates on short-term unsecured construction loans run between 12% and 18% (1.0% to 1.5% per month) depending on individual's borrower credit ratings (Mega Loan Funding 2008).

Table 8-1 shows how retainage affects the actual profit margin in an example $10 million DB project with a target profit margin of 5.5%, over the range of

Table 8-1 Example of Retainage Impact on Profit Margin

Retained Percentage	Total Retainage on $10M Project	Monthly Finance Rate	Months before Release of Retainage	Finance Cost	Actual Project Profit	Actual Profit Margin
0%	$0	NA	NA	NA	$550,000	5.5%
5%	$500,000	1.0%	3	$15,000	$535,000	5.4%
5%	$500,000	1.5%	3	$22,500	$527,500	5.3%
10%	$1,000,000	1.0%	3	$30,000	$520,000	5.2%
10%	$1,000,000	1.5%	3	$45,000	$505,000	5.1%

financing rates. In this example, the owner holds the retainage for three months beyond project completion. Assuming that the design-builder is unwilling to propose on the project if it does not make at least the 5.5% target profit margin, it needs to add $45,000 just to cover the financing cost during the period before release. This does not include the cost of financing the retained earnings during the period leading up to substantial completion. In the extreme case, if the design-builder believes there is a high probability that the retainage will be held indefinitely due to a claim, it would need to add $550,000 to the project cost if it does not want to put its profit at risk.

Options for Retainage in Design-Build Projects

Before including the standard DBB retainage clause in a DB contract, the owner should carefully consider the differences between the two project delivery methods. In traditional low-bid DBB, it is possible—even probable—that the owner may have to award the project to a constructor with which it has no experience and whose qualifications consist only of being able to furnish the necessary bonds and other securities. In DB, the owner has the option to prequalify interested design-builders by using the two-phase award method, with the first phase being an evaluation of qualifications. In the first phase, the owner can require the design-builder to have specific types of qualifications and past experiences, which reduces the owner's risk of awarding to an incompetent constructor. Next, the design-builder's performance on the current project will directly affect its competiveness to win the next project from the same owner. Additionally, other owners that use DB will require references from past DB projects before awarding a contract. Therefore, there is a strong incentive for the design-builder to satisfy the owner's requirements and deliver a high-quality project. One way to get a good recommendation is to complete the punchlist in a timely manner. There is also a disincentive to submit claims on minor issues for fear of "poisoning the well" for future projects. As a result of these differences, the need to hold a substantial amount of retainage as leverage for the design-builder to complete the punchlist is arguably reduced.

There are four options for retainage in DB projects:

- Cease imposing retainage at a point in the project where it appears that the design-builder will be able to satisfactorily complete the project.
- Tie retainage to satisfactory progress.
- Retain a fixed sum for specific purposes.
- Only impose retainage during pay periods where quality issues remain unresolved.

In the first alternative, at the start of the project the owner withholds the stipulated percentage of retained earnings, and it reserves the right to continue retainage throughout the life of the contract. At some point the design-builder's performance demonstrates that it will be able to complete the project to the owner's satisfaction. In other words, the owner is pleased with the design and construction effort, and the project is proceeding according to the agreed-upon schedule. The owner then rewards the design-builder by ceasing to withhold retainage as long as the satisfactory performance is perpetuated. In this option the owner can even choose to begin refunding retainage if it believes this action is warranted. This approach creates a tangible financial incentive to maintain the schedule and comply with contract quality requirements by reducing the amount of carrying costs that will actually be experienced, thus marginally increasing the profit on the project.

The second approach is designed to create an incentive to maintaining the proposed schedule and is best used on schedule-driven projects. Retainage is only imposed if the project is falling behind schedule or has serious quality issues that will put it behind schedule due to the requirement for rework. In this approach, the project's schedule is analyzed in each pay period and this analysis determines whether or not retainage is imposed on that period's progress payment. Additionally, if the project falls behind schedule due to the owner's actions or lack thereof, the contract completion date should be reestablished to make future pay estimates easy to analyze. In this approach the owner could also refund the retainage if the design-builder is able to regain the approved schedule. The cost savings due to refunded retainage may in fact mitigate the cost of crashing critical activities to reestablish the original schedule.

The third option seeks to "make the punishment fit the crime." In this case, retainage would be tied to only those issues that are unresolved. For instance, if during a review a specific design subconsultant's product is found to not comply with the contract, the entire value of that activity in that pay period would be retained until the deficient work is corrected. The same logic applies to trade subcontractors. Rather than "punish" the entire project by retaining a fixed percentage on all work, the owner would designate the work upon which retainage would be imposed, and its amount. This allows the design-builder to fully pay those subcontractors whose work is satisfactory through its standard "pay when paid" subcontract clauses. Again, the owner has the option of refunding all or a portion of the retained earnings when the deficiencies that triggered the imposition of retainage are corrected.

The final method is to make retainage function as an incentive to produce high-quality design and construction. The logic behind this approach requires the output from the project QA program and parallels the logic used in DOT statistical acceptance specifications. For example, the Washington State DOT uses an expected pay curve that is based on their statistical acceptance specification (Muensch and Mahoney 2001). Transferring this idea to retainage would necessitate development of a similar curve, as shown in Fig. 8-3. (Note: This curve is hypothetical and not an example of one that has been successfully applied.) The owner would then take the results of the QA program for the pay period and determine the corresponding percent of retained earnings based on the design-builder's performance during that period. Again, an incentive to achieve high quality is created through the use of retainage.

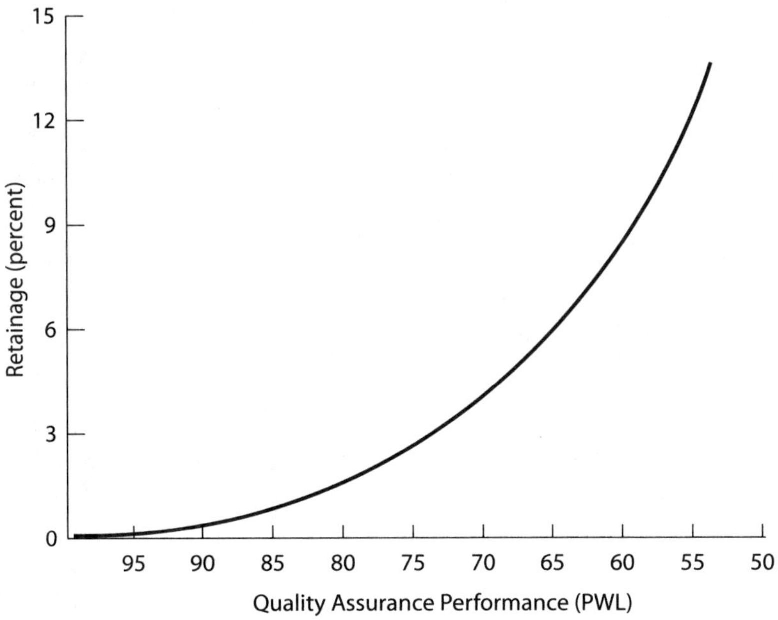

Figure 8-3 Proposed retainage curve based on quality assurance performance. PWL, percent within limits.

No matter how an owner structures its process for applying retainage, it should first evaluate the possible impact on both the DB project's cost and its effect on potential performance.

Final Payment

Final payment in DB projects is not much different from that in traditional construction projects. The major twist is the necessity to conduct a design close-out in conjunction with the construction close-out to ensure that all the required technical information is on hand in case the owner will need it later in the project's operational period. It is recommended that the design close-out be conducted as soon as the completed design is released for construction, to capture all the institutional knowledge of the design team members who will ultimately disperse when their immediate presence is no longer required. The items that need to be collected from the design team and turned over to the owner include:

- Design analyses
- Design calculations
- Geotechnical reports
- Energy studies
- Electrical load studies

- Approved intermediate design submittals
- Final drawings and specifications
- Records of design review meetings
- Other items that would be of use in either the operation and maintenance of the facility or forensically if a failure should occur

The mechanics of the construction close-out will not be discussed here because they are virtually no different from traditional construction projects. However, there are opportunities to improve the process as a result of the internal relationships within the DB team. First, there is no reason why the final inspection of the project cannot be conducted using the completed as-built drawings. The DOR is on the same team as the design-builder and is thus in a better position to assemble these documents; it owns the electronic drawing files and is more intimately familiar with the drawings than the design-builder's field personnel. Additionally, the DOR should have been involved in all field changes or adjustments. Thus, it could return from the close-out meeting and make the necessary changes to the construction documents on an ongoing basis.

Also, the DOR and the design-builder's project manager should conduct a joint pre-final inspection of the project, assemble a punchlist, and correct all the deficiencies on that punchlist before the formal final inspection is scheduled with the owner. Theoretically, the owner should find no deficiencies. In practice, the owner will always find items needing correction, but the list will be much shorter. This will reflect well on the design-builder's reputation for winning future projects with this owner or receiving recommendations from the owner for other owners' DB projects.

Finally, the design-builder should prepare an as-built schedule for the project that includes both the design and construction phases. This will document the project's history and serve as a valuable reference for future projects. Additionally, this furnishes an opportunity for the DB team to review lessons learned and feed those back into their individual business systems to ensure that their value is captured in future projects. This exercise could be conducted with the owner's participation as a final "partnering" meeting to bring closure to the agreement.

Limitation on Payments

All types of DB contracts have a limitation on payments by the owner to the design-builder. The simplest and most common is the firm, fixed price or lump-sum DB contract. The most complex is the design-build-operate-maintain (DBOM) contract where the design-builder must not only propose a price for the capital design and construction effort, but also a long-term, multi-year cash flow system to cover the cost of operating and maintaining the project after it is built. The reason for the limitation is different for public and private projects. In the public sector, capital improvements require some form of legislative authorization and hence there is a need to set a project budget against which the funding can be authorized.

Therefore, public owners must limit payments to literally obey the law by which the project was authorized. In the private sector, a DB project is authorized by some financial analysis predicated on the completed facility's ability to generate revenue that amortizes the design and construction costs over the owner's minimum payback period, and/or generate a required rate of return on the investment of the capital used to deliver the project. To exceed the capital costs or any of the payments anticipated in the project is to reduce the project's rate of return or to extend the period over which the investment is amortized. Thus, the limitation on payment feature of a DB contract is in fact the owner's attempt to quantify its cost risk on the project and allocate it to the design-builder.

There are three fundamental forms for the owner to transfer cost risk through limitation on payments for a DB project:

1. Firm, fixed price or lump sum.
2. Guaranteed maximum price (GMP).
3. Priced cash flow.

However, no matter which of the above methods is used, the rule that must be applied to all of them is that the price upon which the contract is based is inextricably tied to the scope of work for which it was developed. This concept is well understood for the fixed price methods because many DBB contracts use this method to limit payments. Conversely, the term GMP is often misunderstood to mean that the contractor will not be paid above that amount regardless of whether the scope of work increased. This is not what GMP means, and it is patently unreasonable to interpret the term "guaranteed" in this manner. To successfully transfer the cost risk to the design-builder, the owner needs to ensure that the limitation on payment method matches the technical certainty of the project's scope.

Firm, Fixed Price and Lump-Sum Contracts

Regardless of which term is used, firm, fixed price and lump-sum DB contracts require the design-builder to commit to the project's price before design is complete. In this form, the design-builder is at risk for the difference between the as-bid and as-built quantities of work. Thus, it must include a design allowance for quantity growth during design, as well as the usual pricing contingencies for material and labor escalation, and other uncertainties. These types of contracts have historically been used in the DBB building sector because quantity uncertainty is low. However, in the heavy civil and transportation sectors, unit price contracts dominate the DBB field because of the much larger scale of these projects and the inherent variability of quantities in the materials used to construct them. Thus, transferring to a lump-sum price creates a pricing paradigm shift for both owners and design-builders in this industry. Figure 8-4 graphically illustrates the process that a design-builder uses to arrive at a lump-sum price for this form of DB contract.

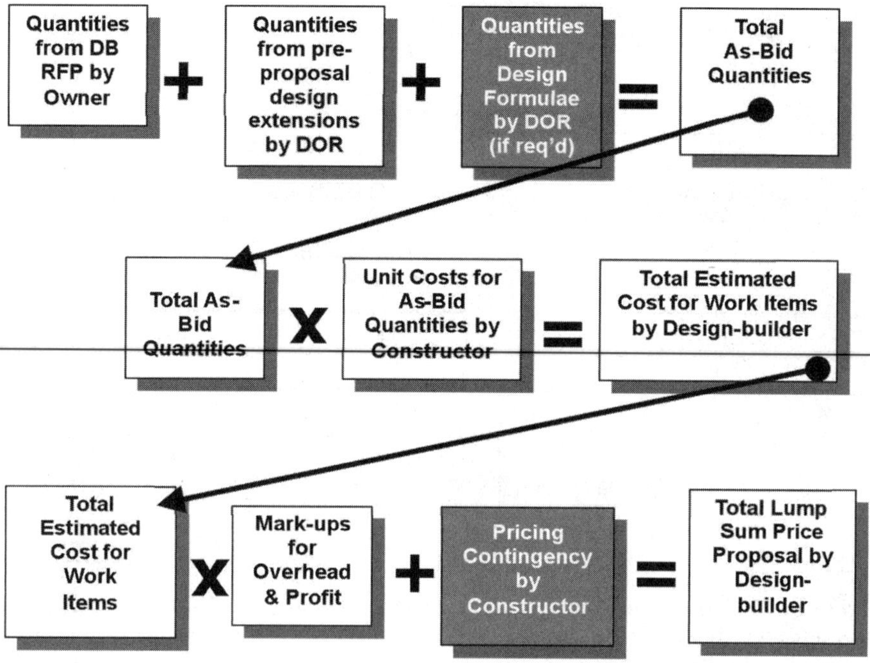

Figure 8-4 DB lump-sum price proposal development. In this figure, Price = Cost + Mark-ups.

The reason why unit price contracts are used in the DBB heavy civil sector is to permit the owner to share the risk of quantity growth with the contractor and thus reduce the project price. In a lump-sum DB contract, the risk of quantity growth during design is added to the risk of quantity growth during construction. Thus, the issue of cost risk is further complicated. A number of owners that build heavy civil projects have dealt with this issue by applying a unit price approach to a select group of pay items where the risk of quantity growth during project execution is unacceptably high. Thus, in this approach, the design-builder will propose a firm, fixed price for the majority of the features of work in the project, and unit prices for the selected high-risk items. This restores equity to the shared risk of quantity change and permits the design-builder to reduce design allowances and construction pricing contingencies for those pay items.

Guaranteed Maximum Price

As mentioned in the introduction to this section, DB GMP contracts create special challenges when it comes to limiting payments. Owners must remember that GMP contracts are used to create a mechanism for shared savings, as opposed to firm, fixed price contracts where all savings go to the design-builder. Figure 8-5 is

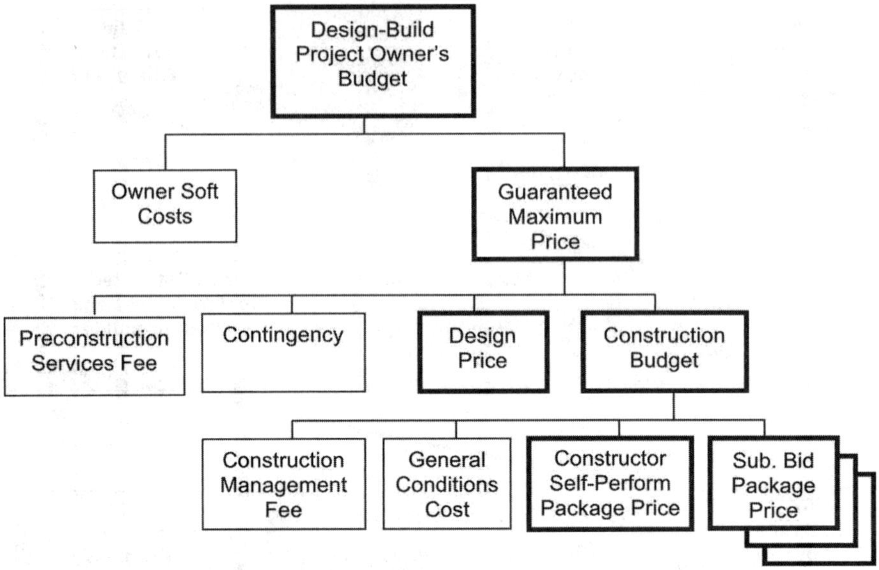

Figure 8-5 DB guaranteed maximum price (GMP) model. Boxes with thin lines indicate soft costs; boxes with heavy lines indicate hard costs. GMP, preconstruction services fee, and contingency are negotiated amounts. Design price, construction management fee, general conditions cost, constructor self-perform package price, and subcontractor bid package price are lump-sum amounts.

a model for a typical DB GMP contract. It breaks the project's budget into hard and soft costs as well as negotiated and lump-sum amounts. The hard costs are those associated with a contract for specific goods or services. For instance, the design budget and the trade subcontractor bid packages are both hard costs because they will emanate from contractual arrangements internal to the DB team. The cost for those features of work that the constructor plans to self-perform are also shown as hard costs because of the need to include material supply and labor contracts in them. The soft costs are those that are less easy to directly quantify and require the application of some professional judgment to ensure that they are both applicable and allocable to the DB GMP. This group includes the constructor's general conditions and all fees and contingencies.

Several of the terms in Fig. 8-5 require definition. First, the term "owner soft costs" is used to cover those costs internal to the owner's organization for project oversight and administration. This may include the cost for a consultant or construction manager to act as the owner's representative during design and construction. Next, "preconstruction services fees" are often not broken out in DB contracts as they are in construction manager-at-risk contracts. However, that does not mean they do not exist. Preconstruction services fees are the costs incurred by the design-builder to coordinate the construction with the design and administer the contract prior to the start of construction. They include the following typical activities:

- Preliminary project evaluations and estimates
- Constructability reviews
- Reviews of reports, studies, and data
- Participation in internal design reviews
- Public affairs activities, if required
- Procurement planning
- Long-lead procurement
- Stimulation of trade subcontractor bidder interests
- Coordination and interaction with subcontractors
- Prequalification of trade subcontractors
- Analysis of supplier bids and proposals
- Trade subcontractor selection and award activities
- Other activities as appropriate

The contingency shown in the figure actually is comprised of at least two parts: the design contingency and the construction contingency. The "design contingency" is assembled from the design allowances the DOR developed for each feature of work for the probable quantity growth during the design process, plus a reasonable amount to cover the DOR's design time growth due to design reviewer comments and other potential design issues such as correction of design errors or omissions. The "construction contingency" is devoted to the typical issues of loss of productivity, increased material and labor prices due to inflation, and extended overhead costs attributed to time growth. Some owners add an "owner's contingency" or "management reserve" to the contingency pool. This type of contingency is for the owner's use to cover unexpected costs or late additions to the scope of work. The "construction management fee" is the builder's profit on the construction itself, and the "general conditions cost" is the builder's indirect or overhead costs during construction.

In the DB GMP model, the limitation of payments becomes the GMP itself. Obviously, a project with no owner's contingency will require the GMP to be adjusted for owner-directed changes to the technical scope of work. If there is an owner's contingency, the GMP will also need to be adjusted when the value of that contingency is exceeded. Under ideal conditions where all pre-project cost estimates are perfectly accurate, the sum of all the contingencies will constitute the amount of savings to be shared in accordance with the contract agreement. In the worst possible case, the design-builder will be responsible for all costs above the GMP that are not attributed to increases in the scope of work or owner-induced delays. Shared savings due to final costs being less than the GMP are explained in detail in the Unused Contingency in Design-Build Projects with a Guaranteed Maximum Price subsection below. Finally, it is important to point out that the use of DB GMP with shared savings requires that the design-builder structure its cost accounting system in a manner that will permit the owner to audit the project's costs and validate the amount of savings to be eventually shared.

Priced Cash Flow

This approach is normally used in conjunction with either multi-year DB projects where the owner has a limitation on available capital it can pay out over time, or in DB projects with postconstruction requirements to operate and/or maintain the completed project for a period of time, such as DBOM and design-build-operate-transfer (DBOT) projects. The first case is usually found in the public sector where the capital funding is subject to fiscal year authorization and the public owner must ensure that it does not commit more money than it is authorized to commit in any given fiscal year. This case is also found in the private sector where a developer's line of credit is capped and it must depend on other revenue to allow it to obtain the necessary financing for the DB project. Thus, the design-builders are required to submit a priced cash flow for every pay period in the project's schedule. These are then used to develop a "maximum payment curve" (Yakowenko 2005). This curve describes the greatest periodic payment the design-builder can request in any given period regardless of actual earned value. In the fiscal year case, these payments would accumulate for 12 months and their total cannot exceed the authorized amount. In this type of project, it is important for the design-builder to analyze its schedule to determine those periods in which it may not be able to be fully reimbursed so that it can accurately figure the carrying costs and add them to the price proposal. It is also important for the owner to evaluate the design-builder's schedule in the context of the cash flow to ensure that it does create a potential dispute by approving a Schedule of Values over time that will exceed its ability to pay.

In DB projects with postconstruction options, the priced cash flow is usually restricted to the operation and maintenance period. The issue in these schemes is to create a meaningful index that can be used to account for the time value of money for extended periods. A number of options are available. Figure 8-6 shows three construction cost indices and how they compare with each other over a 15-year period. The striking thing about this graph is the difference between the local indices and the national-level *Engineering News Record* index. The South Dakota DOT index represents one from a location that is largely rural, whereas the California DOT index more or less represents an urban environment. One can see the volatility at the state level, while the national index is reasonably stable. The lesson learned from this analysis is to find an index that is related to the project instead of one that measures national changes. The index will be used to convert the contract priced cash flow amounts to current year amounts for purposes of making the periodic payments.

Incentive Payments

As stated in the first chapter of this book, most owners select DB project delivery to compress the project delivery period. Hence, it is logical that owners will provide financial incentives for design-builders to deliver that project ahead of the

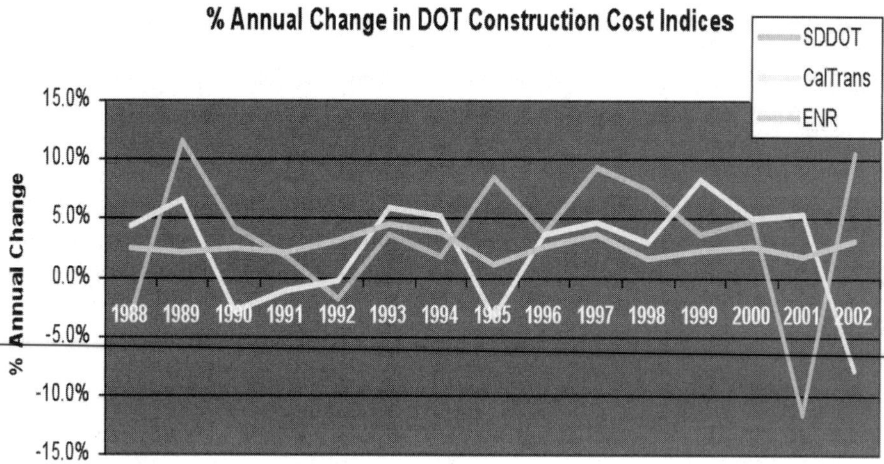

Figure 8-6 Comparison of two state DOT composite construction cost indices and a national construction cost index.

SDDOT, South Dakota Department of Transportation; CalTrans, California Department of Transportation; ENR, *Engineering News Record.*

contract completion date. This is especially true for those projects such as casinos and toll roads that have a revenue stream that starts as soon as the project opens for business. As a result, the most common incentive found in DB projects is the early completion bonus (Yakowenko 2005). However, the need to inspire creativity and innovation in project execution has led owners to devise incentive/disincentive schemes for a variety of project aspects that are particularly valuable to a given owner. The following list shows incentives that have been found in the literature, but is not all-inclusive:

- Early completion bonuses/late completion penalties
- Award fees (community relations incentives/disincentives)
- Quality incentives/disincentives
- Maintenance of traffic and access incentives/disincentives
- Environmental impact reduction incentives
- Shared savings.

Schedule Incentives

Early completion incentives typically base the financial amount on the number of days earlier than the contract completion date that the design-builder achieves substantial completion (Quatman 2000). At times these incentives can be quite large if the benefit of early opening justifies it to the owner. The recent collapse of the I-35 bridge in Minneapolis created a situation where the Minnesota DOT (MnDOT) needed to structure the DB contract in a manner that got the new bridge open to traffic as quickly as possible. MnDOT calculated that it was costing the state $400,000 per day in user congestion costs while the bridge was out of operation. The DB contract offered a $200,000 per day early completion bonus, up to a total of $7 million, on the $234 million reconstruction project (Kessler 2008). Another DB project with an incentive was the Oakland Bay Bridge Ramp Reconstruction project that occurred in 2007 when a fuel tanker truck exploded on the ramp. This project was expected to take 50 days, with a $200,000 per day early completion bonus not to exceed $5 million. The winner completed it in just 17 days and collected the entire bonus (Gingrich 2007). Most schedules come with a comparable disincentive for finishing late. This is certainly fair and, as the two above examples show, the DB industry is definitely influenced by the presence of these contract features.

Award Fees

Award fees are incentive payments made periodically throughout the project "for performance that exceeds predetermined levels of performance that are set in the contract requirements" (Yakowenko 2005). The theory of award fees is to incentivize key elements of performance and reward the design-builder if it delivers more than the contractual minimum. The Utah Transit Authority uses these types of incentives on its DB projects (Touran et al. 2008). This owner assembles a standing committee of third-party stakeholders such as utility companies, affected municipalities, and the state environmental agency, that meet every pay period to determine how much of the award fee (typically 2% to 3% of the periodic pay request) the design-builder has earned. They assess performance in areas such as traffic congestion, timely public information, unscheduled utility outages, noise and dust pollution, etc. Essentially, this program puts about half the design-builder's profit at risk in each pay period. The program has been quite successful and has encouraged the design-builders in a manner that not only resulted in timely completion but also minimized the impact on the communities in which the projects were constructed. The Minnesota DOT offered a $100,000 incentive to the design-builder on its ROC 52 project for exceeding the standards for public relations (Gladke 2006).

Quality Incentives/Disincentives

A Transportation Research Board report on quality in DB projects lists the motivation for quality incentives as follows:

> There are reasons for the use of incentives. Incentives provide added value, going beyond the minimum contract requirements and enhancing the product to perform better under service; are quantifiable, on the basis of measured data or test results specifically related to goals; are achievable, able to produce a uniform product consistently; provide extra money for product enhancement; and are good for public recognition. (Gladke 2006)

Figure 8-7 is an example of a quality incentives/disincentive program instituted by the Washington State DOT. In that figure, one can see that the builder can earn a maximum 5% bonus or lose up to 25% of its pay for specific pay items if the QC tests fall within specified limits. Other public transportation agencies have instituted these systems for pavement smoothness and other construction features. In the architectural sector, incentives have been offered for design-builders being able to obtain certain building permits without delay. This may seem like a schedule incentive; however, it is really an incentive to enhance design quality in a manner that meets or exceeds the applicable code requirements for the specific permit.

Traffic and Access Incentives/Disincentives

Reducing congestion and ensuring that the traveling public is able travel through a DB project is the paramount factor in transportation infrastructure projects. User costs of congestion due to delays, increased toxic emissions, and other factors on urban freeways can exceed $10,000 per lane-mile/day (Hicks and Epps 2000). Thus, a typical eight-lane urban freeway would incur a user cost of $80,000 per mile per day, and a hypothetical five-mile DB urban freeway upgrade project would generate user costs on the order of $400,000/day. Taking the analogy one step further would lead one to determine that obstructing traffic for a 30-day period during the project's construction costs commuters $12 million on that five-mile stretch of freeway. If the design-builder is able to manage the project in a manner that allows the passage of the critical volume of traffic, it generates a

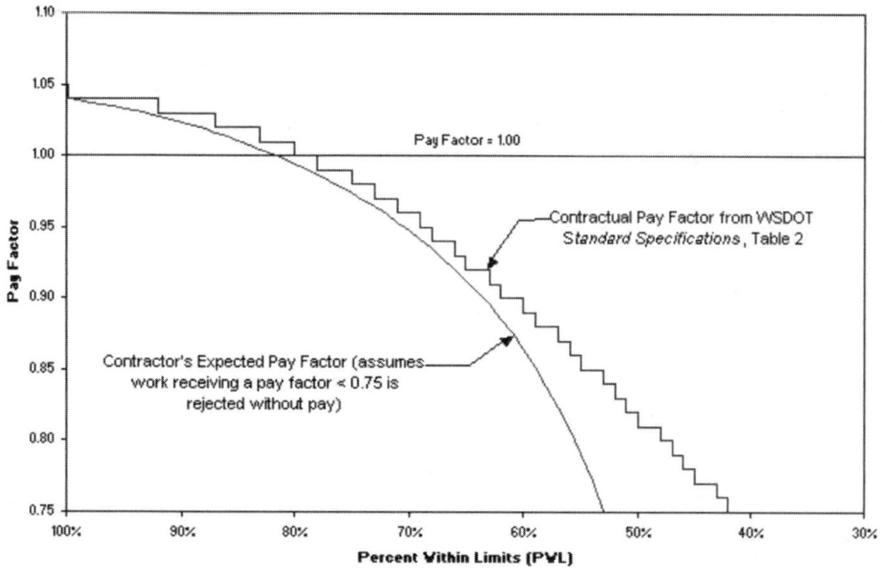

Figure 8-7 Washington State DOT Expected Pay Curve Based on Quality Assurance Program Output (Muensch and Mahoney 2001)

significant benefit and hence it is probably worth incentivizing to encourage innovative solutions to the maintenance of traffic problem.

The Arizona and Utah DOTs both include an incentive payment for the design-builder furnishing a means to sustain traffic flows in excess of the stipulated minimum requirement. Both agencies use intelligent transportation system technology to measure actual traffic volumes through DB project work zones. Additionally, access by landowners to properties adjacent to the project also carries an incentive if the design-builder is able to use less than a set number of allowable obstructed access days. The incentives are used to encourage the DB team to think out of the box and to find ways to overcome these critical public relations issues.

Environmental Impact Reduction Incentives

Perhaps one of the fastest growing incentive categories in construction revolves around encouraging designers and constructors to decrease the project's impact on the environment. Incentives in this category come in various packages. First, municipal building officials and state governments offer outright grants for owners to build "green" buildings. These can be used in the DB context to offset the owner's cost of the DB contract itself. However, if that is the case, the owner will need to include language that requires the design-builder to deliver a facility that qualifies under the special rules of each program. Design submittal reviews will also need to be made with these grant requirements in mind to ensure that the final design will be found eligible for the grant. In these cases it is also advisable to invite a representative from the granting agency to attend the initial design meeting to assist in sorting out any issues that might arise.

A second type of environmental incentive seeks to encourage the design-builder to exceed the minimums required in the applicable environmental permits. For example, the Maryland State Highway Administration included an incentive that paid the design-builder $35,000 per acre for each acre of reduced wetland impacts below the amount allowed by the Section 404 permit (Louis Berger Group 2005). On this project, the design-builder was able to reduce the wetland impact by 25%, amply demonstrating the success of such an incentive scheme.

The final type of environmental incentive comes from the building sector. The U.S. Green Building Council (USGBC) is the proponent for the Leadership in Energy and Environmental Design (LEED) Green Building Rating System™. This system awards points for various sustainable design and construction features. The system has a total of 69 points available, and a building can achieve one of the following four levels of certification:

- Certified: 26–32 points
- Silver: 33–38 points
- Gold: 39–51 points
- Platinum: 52–69 points

An increasing number of public and private building owners are now requiring that their projects achieve some level of certification. The environmental incentive in these projects is tied to exceeding the required minimum certification. For example, if a DB building project was required to achieve LEED Silver certification and, at the end of the project it scored 45 points, the incentive payment would be calculated on the basis of the 7 points by which the design-builder exceeded the 38 points on the top end of Silver certification. In fact, this system can be used to create an incentive for a project that will not achieve LEED certification by merely specifying a minimum number of LEED points and paying a bonus for every point above that number. Thus, the owner can convey its desire to be as environmentally friendly as the budget will allow, even if the project does not receive certification upon completion.

Shared Savings

There are a number of ways that shared savings is used as an incentive in DB contracting. All are intended to create a situation where both the owner and the design-builder can benefit from the design-builder's innovation and first-rate design and construction management. Additionally, these schemes can be used to reduce the design-builder's contingencies related to the risk of committing to a firm, fixed price before the project's design is complete.

Value Engineering in Design-Build

The most common shared savings scheme is to include a value engineering (VE) clause in the contract to encourage the design-builder to seek ways to reduce project cost while furnishing the same level of performance and service. The use of VE in DB is somewhat controversial. On one side of the argument are those who maintain that all the VE is done during the proposal preparation phase and that including a VE clause merely creates a bias toward deliberately overdesigning so that an extra profit margin can be realized through the VE incentive clause. The other side argues that VE is the owner's "last bite at the apple" and furnishes it with an opportunity add value to the project by leveraging the special knowledge and skills of the specialty design subconsultants and the trade subcontractors.

The two sides of this argument can be easily reconciled by clearly defining the technical baseline against which VE proposals can be measured. That baseline should be the technical content of the DB RFP and any other owner-developed or -prescribed design upon which the winning proposal was based. It is always possible that the owner has either intentionally or unintentionally excluded potential design solutions to a given design problem in its solicitation. The VE clause furnishes a contractual mechanism to revisit those decisions and, as in DBB VE clauses, the owner always has the right to disapprove the proposed change. Additionally, the VE clause furnishes a protection against a protest from an unsuccessful offeror if the owner chooses to implement a design-builder's proposed VE change and that change fundamentally alters the scope of work. The reason is that all the competitors proposed on the project knowing that it included the potential for VE changes. If the VE clause is not contained in the DB RFP, then the protester has a strong legal ground to delay or halt the contract's progress.

The second issue with VE incentives is whether to allow the design-builder to VE its own design product after the owner has reviewed and released it for construction. In most cases this is not appropriate. However, the "last bite at the apple" theory would argue for allowing this. This can be handled in one of two ways. First, the owner can take the position that the design-builder made an error in its first design by not including the proposed betterment in the initial design submittal. Then the owner will allow the change to be made and take full credit for any savings. The other option is to include a second level of shared savings for VE of the design-builder-produced design that creates an incentive for the DOR's design team to keep searching for more economical solutions throughout the project. For example, the shared savings for the VE of the owner's product could be a 50-50 split, whereas the savings would be shared at a 90-10 ratio in favor of the owner for VE of the design-builder's product. Regardless of how the VE clause is written, it seems imperative that there should be one in all DB project contracts.

Unused Contingency in Design-Build Projects with a Guaranteed Maximum Price

Pertti Lahdenperä, writing for the Technical Research Centre of Finland, said:

> Efficient implementation of a project, with good management by the design-builder and timely support from the owner, creates a savings pool as regards GMP that should benefit both the design-builder and the owner. Most GMP contracts address the issue of allowing shared savings to create an incentive for the design-builder to save costs. (Lahdenperä 2001)

The objective of using a GMP approach to DB project delivery is to reduce the contingency that the design-builder must necessarily include in its lump-sum proposal to cover the risk due to the incomplete design at the time of contract award (Lahdenperä 2001). With a GMP approach, the design-builder can advance the design to a point where it is more confident in its cost estimate; this creates a savings to the owner in reduced contingency costs. By making the remaining contingency visible as part of the GMP, it creates a mechanism to compute final costs and a built-in incentive if savings are either shared or assigned to the design-builder. Figure 8-8 illustrates the concept of how this incentive scheme operates.

The Design-Build Institute of America (DBIA) is the proponent for a DB model cost-plus contract form that uses a GMP negotiated after contract award rather than a lump-sum price that is set before contract award. In this approach, the design-builder is required to include a stipulated contingency in the GMP. The contingency is ". . . for the Design-Builder's exclusive use, even to correct negligent errors. The Design-Builder notifies the Owner of charges against the contingency, but it is not available to the Owner for any reason" (Friedlander 2003, p. 17).

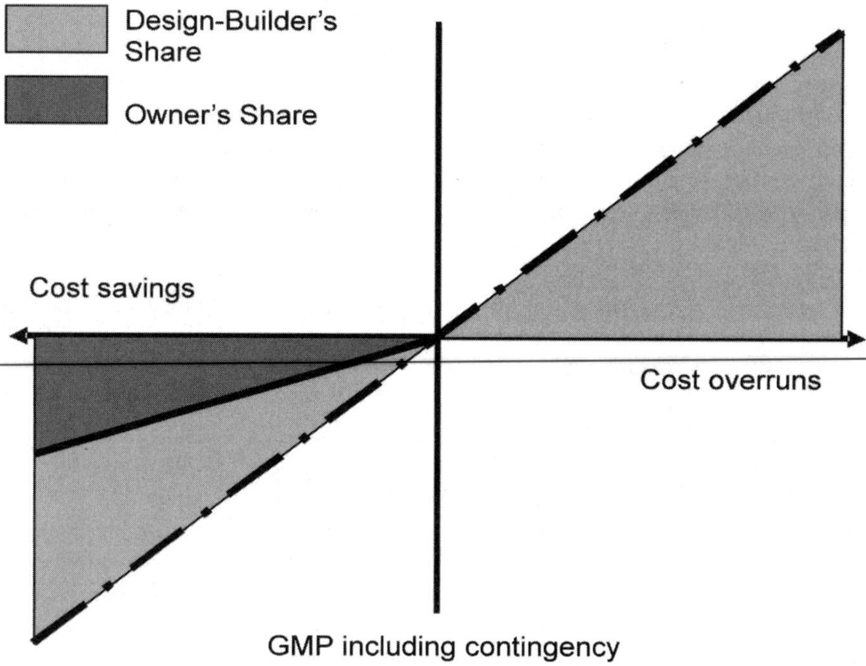

Figure 8-8 Shared savings of the GMP contingency concept.
Source: Adapted from Lahdenperä (2001), Figure 12.

Thus, the design-builder has a contractual incentive to complete the project for less than the GMP and pocket the unused contingency. From the owner's perspective, this makes good sense in that the project is being paid for on a cost-plus basis. Therefore, creating an incentive to contain costs to their lowest possible amount gives the owner the benefit of enhanced cost certainty in a project where cost certainty is inherently low.

Some lump-sum DB contracts also include a visible contingency, but often the unused contingency is refunded to the owner, thus removing the incentive to the design-builder (Friedlander 2003). On the other hand, a 2003 study of eight state DOT DB programs found that at least one (the report does not identify the source) uses the unused contingency as an incentive, sharing the savings with the design-builder (CTC and Associates, LLC 2003). This approach will require an "open-books" style of accounting for the available contingency. Most contingencies are established incrementally for specific risks, and then the risk contingencies are added together to create the project's contingency pool. To implement this incentive scheme the DB contract will require verbiage as to exactly who and under what circumstances can allocate a portion of the contingency for a given reason. To counter potential conflicts of contingency ownership, some projects have broken the total pool up into an owner's contingency, a design-builder's contingency, and a shared contingency (sometimes called the "management reserve") (Lahdenperä

2001). Thus, when the owner needs to add scope to the project, it has a designated pool of funds that it can use at its discretion as long as the pool is not empty. The same applies to the design-builder, who will use its contingency to cover quantity growth during design, escalation of material or labor costs, and other typical reasons. Finally, when one or both contingency pools are emptied, the shared contingency is utilized with joint agreement. This extra pool can also be tapped by joint agreement to cover unexpected costs that neither party could anticipate, such as a change in building or environmental codes.

The use of a "progressive GMP" or "incremental GMP" in DB projects is starting to emerge as a variation on the single GMP theme. Under this approach, the DB project is broken up into phases or work packages. An incremental GMP is negotiated at the completion of the design of each package and, when the final package is complete, the final GMP is calculated as the sum of the incremental work package GMPs. Again, the theory here is to reduce the design-builder's risk of cost growth during design and force contingencies down to their lowest possible level. One advantage of this method is that work packages that include costly materials whose price is volatile, such as structural steel or asphalt, can be designed and released for construction early in the schedule, allowing the design-builder to lock-in their pricing at the earliest opportunity. The shared savings incentive can be applied to progressive GMP DB projects in the same manners as described above. Intuitively, the total amount of the contingency should be less than in single GMP projects, and therefore the total incentive will be lower.

Design-Build Project Cost Control

Controlling costs during DB project execution guarantees both the owner's budget and the design-builder's profit margin. Therefore, it is important to keep cost control in mind while completing the design and construction tasks. The DOR's design team is the entity in the DB project that must make the most dramatic shift in its practices to ensure that the as-built costs do not exceed the as-bid costs. In traditional DBB, it is not uncommon to find that cost estimates are not completed until the design is nearly complete. Design professionals typically focus on technical requirements first, with cost constraints occupying a secondary position of importance. Additionally, it is also common to find that the DOR's cost estimate is different from the prices provided by construction contractors in their bids. If the bids exceeded the owner's budget, then the DOR was required to VE the design back into the available funding. Thus, including construction cost control as an integral aspect of the design process is a critical project delivery factor that is missing in many design professionals' routine practices. Friedlander (2003) puts the issue in this light:

> In designing a project, architects and engineers often do not take into account such issues as ease of constructability and availability of materials. Although a design professional may be contractually obligated to design to a

> fixed limit of construction cost, often the A/E does not have the necessary construction information to ensure that the bids on the contract documents will be within the budget. . . . Furthermore, even detailed construction estimates developed by consultants during the design phase may be significantly inaccurate if market conditions are different when actual bids are solicited.

This is not to say that construction professionals do not need to adjust their DBB practices when entering into a DB contract. The major shift for constructors is the requirement to mitigate the impact of design errors and omissions. In DBB, these would have been attributed to the owner and provided the design-builder with an opportunity to increase its profit margin because change orders are sole-source procurement actions. Thus, in DB it is incumbent upon the construction team to participate in the design process and help the DOR to flush out errors, omissions, ambiguities, and inconsistencies in the construction documents. It is also imperative that the constructors report design issues as soon as they are identified so that the DOR's team can address and clarify or correct them. Therefore, close cooperation between all members of the DB team is the primary tool for controlling costs post-award and during project execution.

Design Cost Control

There are two facets to the design cost control problem. The first is controlling the cost of the actual design work itself and ensuring that the actual cost is less than or equal to the design budget in the DB price proposal. One of the primary threats to design costs is scope creep. This is an insidious process where numerous small and seemingly insignificant changes and/or improvements are made to the basic design by the designers themselves, trying to make the facility as good as possible. When these are added up, the required design effort is greater than that contemplated in the bid. Scope creep can also occur as a result of design reviewer comments that actually accomplish the same result while the design professionals strive to satisfy the owner's desires or preferences without checking to see whether the comments are in fact a change in scope. Kuprenas and Nasr (2003) put it this way:

> No matter how it occurs, scope creep is detrimental to the overall project budget and schedule. It results in additional design phase charges due to extra design work. In addition, it results in construction cost and schedule overruns due to increased construction scope.

Those authors proposed the development of a metric based on earned value theory that they call the "design cost performance index" (DCPI) (Kuprenas and Nasr 2003). The equation for this metric is:

$$\mathrm{DCPI} = \frac{\mathrm{BCDWP}}{\mathrm{ACDWP}} \tag{8-1}$$

where BCDWP is the budgeted cost for design work performed in a given period and ACDWP is actual cost of design work performed in the same period.

From Eq. 8-1 it can be seen that if the DCPI equals unity, then the design work is precisely on budget and schedule. If the DCPI drops below 1, the actual costs are exceeding the budget and the design project manager must take action to determine the cause before the opportunity to regain the budget is lost. The beauty of this metric is in its simplicity. Most design firms require their personnel to charge their time to specific projects against which a budget based on billable hours was developed. Thus, if those payroll expenses are accumulated on a periodic basis and compared with the plan for the period, the design manager can very quickly and easily control design costs.

Design Allowances

The second aspect of cost control during design is ensuring that the as-designed cost of the DB project itself does not exceed the as-bid cost. Becker (1990) states that "[A] critical element for controlling the cost, schedule and scope of a project is gaining and maintaining control of the design process. Failure to control and manage this process will result in delays and increased construction costs." Controlling the design process means ensuring that the final design conforms to the constraints formed by the assumptions used to develop the proposal. The "new sense of discipline" is displayed by ensuring that the final as-designed quantities of work do not exceed the quantities based on the design information that the DOR gave the constructor to develop the estimate. This is done through consciously and repetitively looking at design alternatives that meet the owner's design criteria to determine which ones best achieve the budget, schedule, and constructability constraints established in the proposal. One author describes this "design-to-cost" process rather succinctly:

> After preliminary cost-performance tradeoff analyses are used to set initial performance and cost objectives, cost is established as a constraint rather than in the traditional manner as a dependent variable. (Wollover 2006)

Putting this statement into a DB context, this author advocates making the cost performance trade-off decisions during proposal preparation and then completing the design within the scope described by the budget constraints. This would mandate that the DOR promulgate this attitude all the way to the bottom of its organization, as well as among its design subconsultants, to ensure that every design professional understands that the design solution is constrained by the limits established in the price proposal.

This statement in no way argues that the budget is preeminent over the design criteria and performance requirements set by the owner in the DB contract. These must be met regardless of the ultimate impact on the budget. It does argue that all the design professionals who will make input to the final design be made fully aware of the time and cost constraints that are apparent in the project, and that they regulate their performance in such a fashion as to ensure that these constraints are not violated. If they are, the designers have a duty to seek acceptable design solutions that mitigate the impact to the budget.

In light of the above discussion, it can be concluded that the DOR in a DB project has a duty to ensure that the scope of the final design does not exceed the constraints on cost, schedule, and constructability that were set forth in the winning proposal. The best way to ensure that this important objective can be achieved is to ensure that the estimated as-designed quantities of work are based on reliable parametric design formulae or pre-proposal detailed designs to furnish the constructor with information upon which to base an estimate with an acceptable amount of uncertainty. If necessary, a design allowance for each item of work that has not been designed to the point where its material quantities can accurately be determined is added for quantity growth during design. Thus, the designers should have a quantity budget that relates to the feature of work that they are designing, and then each designer should periodically check the as-designed quantities to verify that they have not crept past the quantity budget. If this has happened, then each designer should first seek to reduce those quantities by any technically acceptable means, such as reducing an overly conservative factor of safety if that can be done. If not, the designer must notify the design manager as soon as possible to give the maximum amount of time for that individual to seek potential solutions by cross-leveling the design contingency between disciplines.

Construction Cost Control

Construction cost control in DB mirrors the process that most sophisticated constructors use in their DBB projects. The major cost control issue is to identify design conflicts and errors as quickly as possible to give the DB team the maximum amount of time to correct them and thereby mitigate their impact on the project's budget and schedule. Controlling construction costs requires rich, continuous communications between the members of the DB team and the owner. J. E. Fish (1991) commented, "Managing the design/build process requires defining what is needed, how to obtain what is needed, who should participate, and the monitoring and control necessary to verify that construction conforms to specifications." Thus, DB construction cost control must have a strong linkage to the construction quality management system.

To summarize the discussion in this chapter, payment in DB projects is a complex task due to both the speed at which a DB project can be executed and the parallel nature of design and construction, which in some cases removes the traditional benchmarks that owners use to identify satisfactory progress for payment purposes. Additionally, because of the accelerated delivery period, strong cash flow during both design and construction is imperative in order to gain and sustain the necessary momentum to achieve aggressive completion dates. At the same time, the owner must still protect itself against potential default, even though the qualifications-based selection of both designers and constructors has reduced the probability of that contractual catastrophe. Therefore, owners must review the process that is used to make payments in traditional design and construction contracts and amend it to fit the demands of the DB environment. Even the most financially robust design-builders can withstand only so much cash flow constraint during project delivery before they must reduce the pace of work to pay for labor, materials, equipment, and subcontractors. Additionally, the owner's future reputation with regard to its payment practices will inevitably be transferred into the pricing of the next DB project to cover either increased or decreased carrying costs included in the proposal.

In line with the idea of adjusting traditional payment processes, there are a number of "out of the box" ideas to address this shift in philosophy. First, owners must remember that there is no such thing as a "cost of doing business." Design-builders must recover *all* the costs generated by the DB project to realize a profit. One major item that generates carrying costs in all projects is the cost to mobilize and complete the first period of work to be able to request the first partial payment. In some DB projects in remote locations, that cost can be considerable; if the owner has not accurately estimated its costs (including the cost of financing the first couple of pay periods), this can have a significant impact on the prices proposed by the competitors for the project. One way to reduce this cost (which, incidentally, does not add value to the project in that it is purely an administrative cost) is to offer an up-front payment for mobilization before it takes place. Thus, the owner is using project funds to essentially give the design-builder an interest-free loan to reduce the amount of carrying costs in the project. One owner who did this then reduced the next 12 progress payments by 1/12 of the up-front payment, which literally turned it into a 12-month interest-free loan. Reimbursing certified procurement costs has been previously mentioned in this chapter, and doing this also reduces a project's carrying costs. Research has shown that incentives of all kinds positively affect the way design-builders price DB projects. However, creating an incentive to design and build to minimize life-cycle costs (not just capital costs) reap benefits for the life of the project.

Finally, the details of an owner's DB payment system should be clear and unambiguous. They should be known and understood before the design-builder prices the project, and they should be followed as the project progresses. To finish with the thought expressed by Fish (1991), cost control in DB is purely an exercise in communications and oversight. Design-builders will either be in "maximize profit mode" or "minimize loss mode." Practical experience shows that those design-builders that are making a profit during project execution deliver a better product. Keeping the design-builder in that mode demands a steady flow of payments that represent the value of the project as it develops. Returning to DBB payment policies has the potential to reduce or even eliminate some of the benefits the owner wanted to accrue by selecting DB project delivery.

References

Bausman, D. (2005). "Retainage practice in the construction industry." *Foundation of the American Subcontractors Association, Inc.*, <https://www.asaonline.com/eweb/DynamicPage.aspx?Site=ASA&WebKey=13d42386-3442-4790-bb72-7301acfe73e2&ListSearchFor=Retainage%20practice> [March 14, 2010].

Becker D. (1990). "Controlling construction cost during design." *1990 AACE Trans.*, AACE International, Morgantown, W.V.

Collier, P., Corbett, M., and Lundrigan, B. (2001). "Quality performance in a design-build mega-project." *Proc., 55th Annual Quality Congress*, American Society for Quality, Milwaukee, Wisc., 17–26.

CTC and Associates, LLC. (2003). "Implementing design-build." *Transportation Synthesis Report*, Wisconsin Department of Transportation, Madison, Wisc.

Fish, J. E. (1991). "Cost control in design/build." *Cost Eng., J. AACE Int.*, 33(10), 7.

Friedlander, M. C. (2003). *Risk allocation in design-build construction projects*, Schiff Hardin, LLP, Chicago, Ill.

Gingrich, N. (2007). "The Minneapolis bridge, Katrina and the world that fails." *Human Events*, <http://www.humanevents.com/article.php?id=21841> [July 20, 2008].

Gladke, J. (2006). "Incentives on design-build projects: Two different approaches." *Design-Build: A Quality Process Transportation Research E-Circular No. E-C090*, Transportation Research Board of the National Academies, Washington, D.C.

Gransberg, D. D., Molenaar, K. R., and Datin, J. N. (2008). "Quality assurance in design-build projects." *National Cooperative Highway Research Program Synthesis 376*, Transportation Research Board of the National Academies, Washington, D.C.

Gransberg, D. D., Koch, J. E., and Molenaar, K. R. (2006). *Preparing for design-build projects: A primer for owners, engineers, and contractors*, ASCE Press, Reston, Va.

Hicks, R. G., and Epps, J. A. (2000). "Life cycle cost analysis of asphalt-rubber paving materials." *Proc., Asphalt Rubber–2000*, Rubber Pavements Association, Tempe, Ariz., 569–590.

Holland, J. K. (2002). "Contract documents of the Design-Build Institute of America." *International Risk Management Institute*, <http://www.irmi.com/Expert/Articles/2002/Holland05.aspx> [July 19, 2008].

Kessler, P. (2008). "Reality check: Speedy road construction," *WCCO Television*, <http://wcco.com/local/speedy.road.construction.2.735163.html> [July 20, 2008].

Kuprenas, J. A., and Nasr, E. B. (2003). "Controlling design-phase scope creep." 2003 *AACE Trans.*, AACE International, Morgantown, W.V.

Lahdenperä, P. (2001). *Design-build procedures: Introduction, illustration and comparison of U.S. modes.* Technical Research Centre of Finland, VTT Publications, Espoo, Finland, 452.

Louis Berger Group. (2005). "Design-build environmental compliance process and level of detail: Eight case studies." AASHTO Standing Committee on the Environment, American Association of State Highway and Transportation Officials, Washington, D.C.

Mega Loan Funding. (2008). "Construction loans." *Mega Loan Funding*, <http://www.megaloanfunding.com/construction.htm> [July 20, 2008].

Muench, S. T., and Mahoney, J. P. (2001). "A quantification and evaluation of WSDOT's Hot Mix Asphalt Concrete Statistical Acceptance Specification." *Research Report WA-RD 517.1*, Washington State Department of Transportation, Transportation Center (TRAC), Seattle, Wash.

Office of Program Policy Analysis and Government Accountability (OPPAGA). (2000). "Inflexibility in contracting and retainage practices could hurt construction industry."

OPPAGA Review, Office of Program Policy Analysis and Government Accountability, State of Florida, Tallahassee, Fla.

Quatman, G. W. (2000). *Design-build for the design professional,* Wolter Kluwer Law & Business/Aspen Publishers, New York.

Saltz, S. J. (2007). "The new AIA design-build contract from the design-builder's perspective." *Probate and Property Magazine,* American Bar Association, Washington, D.C.

Touran, A., Gransberg, D. D., and Molenaar, K. R. (2008). "A guidebook for the evaluation of project delivery methods." *Transit Cooperative Research Program Report G-08,* Transportation Research Board of the National Academies, Washington, D.C.

United States Agency for International Development (USAID). (2001). *Requests for qualifications (RFQ) No. 294-2001-004 Jenin–Nablus Highway design-build project, Section H-8,* USAID Mission to the West Bank and Gaza, Tel Aviv, Israel.

Weaver, S. C., and Weston, J. F. (2001). *Finance and accounting for nonfinancial managers,* McGraw-Hill Professional, New York.

Wollover, D. (2006). *Design-to-cost decision support system,* Coleman Research Corp., Arlington, Va.

Yakowenko, G. (2005). "Current design-build practices for transportation projects: Developed and Maintained by the AASHTO Joint Task Force on Design-Build." *USDOT Federal Highway Administration,* <http://designbuild.transportation.org/?siteid=63&pageid=1226> [July 13, 2008].

NINE

Construction Administration Case Studies

This chapter builds on the case studies detailed in Chapter 5 by analyzing the fundamental issues that occur in construction of a DB project, through the presentation, analysis, and discussion of four case studies. These studies illustrate various aspects of the process and the associated challenges. The cases range from disagreements that ensue when one "assumes" a design approach, to the ramifications of not letting the owner's staff understand changes in quality processes. Once again, the cases are drawn from real projects and have been sanitized to prevent identification of the actual participants. The format for each case covers the following: first, the facts and situation; second, a discussion of the issues and positions; third, the outcome; and fourth, lessons learned.

Case 9-1: "Trust Me. I Know This Will Work."

The owner had designed and built many buildings on its headquarters campus and usually required basements to enhance the storage capacity of each structure. It also preferred to store high-value testing and laboratory equipment in the basements to enhance the security of those items. Because of the value of this equipment and its sensitivity to water, the owner was willing to invest in ensuring the integrity of a waterproof basement, and this was considered a key element to project success. The water table was generally high in that area and past experience dictated the use of a mud slab with a waterproof membrane beneath the slab and on the basement walls to adequately waterproof the basement. The level of design in the DB RFP contained minimal drawings but it did contain detailed performance specifications. The RFP stated: "The basement will be designed and constructed to be waterproof for typical conditions in this area." When the performance specifications were written, the owner was expecting a mud slab.

The project was a fast-track project to deliver a vitally needed facility. To facilitate the schedule, the owner agreed to approve the design incrementally, using over-the-shoulder reviews. When this issue arose, the design-builder had completed the footing design, had it approved, and had begun construction. Before the footing excavation began, the excavation subcontractor placed a gravel base in the basement that the owner's field representative assumed was preparation for a mud slab. Later, the owner discovered that the gravel was only a working surface. Additionally, the top of footing elevations used while setting the formwork would not allow room for a mud slab.

When asked about the waterproofing design, the designer-of-record (DOR) verbally described the technology contemplated in the proposal based on the RFP's performance criteria. Drawings were being developed concurrently at that time. Based on past experience with waterproofing basements on this type of base, the owner's representative recommended that the footing construction be stopped to mitigate cost. The design-builder assured the owner that the planned system would work based on the past experience of the designer and the constructor. They reached an impasse where the design-builder stated that if the owner wanted a mud slab, that would constitute a change in scope, and requested a change order and a time extension if directed to change to the mud slab technology.

The two issues at hand in this case were:

- If the owner directed the design-builder to cease work and change the waterproofing design, was the design-builder authorized to issue a compensable a change order?
- If a change order was required, did the owner then assume design liability for the performance of the directed mud slab waterproofing system?

Outcome

This case illustrates a variation on a common problem in DB project delivery. The owner's technical staff will often "assume a design" and then write the DB solicitation documents around it, eliminating other innovative potential solutions. Here, the owner assumed that the design-builder would "know" that past projects included mud slab waterproofing and would include it in this one as well, to satisfy the requirement to design for "typical conditions in the area." As a result, it failed to identify its preferred basement waterproofing design in the RFP. The design-builder was presumed to be competent in designing and installing an "adequate" system for this facility, as evidenced by the award of this contract. Therefore, the owner would have pay for a directed change to the scope of work if it insisted on the mud slab design solution, and the owner would assume liability for its ultimate performance if it was designed and installed properly by the design-builder. Additionally, the owner would have to grant a time extension and pay for redesign and any rework.

In this case, because the schedule was the preeminent factor in selecting DB on this project, the owner ultimately decided, to allow the DB contractor to proceed as originally planned. Interestingly, when we collected the facts of this case, the project had been in use for several years and basement leakage was not a problem.

Lessons Learned

Every sophisticated owner that has an active, ongoing capital improvements program will have design preferences for various features of work based on past experience and the owner's ability to tolerate the risk of trying something new. In this case, if the owner had included the requirement for the mud slab waterproofing in the RFP, there would have been no dispute. Another common problem in DB contracting is professional differences of opinion among design or construction professionals about the adequacy of a given technical feature. These must be resolved based on the technical requirements of the contract. If the contract does not mandate or exclude a specific technical approach, the design-builder can reasonably use it in its design and/or construction deliverables. Therefore, this case graphically illustrates the need for the owner to articulate its technical preferences before the DB contract is awarded.

Case 9-2: "The Winning Inspector Is . . ."

A state DOT decided to try the DB project delivery method and wanted to shift many of the owner's traditional roles to the design-builder due to an acute shortage of qualified engineers on its staff. Quality control was one of the roles it chose to transfer. Because the design-builder was responsible for both design and construction, the agency decided to ensure the final quality through the use of extended warranties on critical features of work. The following two clauses outlined this transfer of risk in the owner's RFP:

- Product warranties will be used to the maximum extent possible to ensure project quality. Because many of the quality assurance/quality control processes traditionally done by XXDOT are being transferred to the design-builder, warranties will act as a means to ensure that high quality standards are being met.
- XXDOT will maintain an oversight role during design and construction in a manner that satisfies federal quality assurance requirements.

This DB project involved a concrete bridge deck. The owner had transferred the quality control for the reinforcement placement to the design-builder and asked for an extended bridge deck warranty. The owner retained responsibility for approval of materials through a quality assurance sampling process. While collecting samples of material, one of the owner's inspectors observed dowel bar and tie bar placement in the bridge deck that she perceived to be noncompliant with traditional standards. The inspector informed the design-builder's QC supervisor on the site and was told that the placement met their internally approved shop drawings. The owner's inspector then directed the design-builder to stop work and halt the concrete pour until the perceived problem was corrected.

The issues at stake in this case were:

- Did the owner have the right to stop work?
- If the owner stopped work and the placement met the standards, was the design-builder entitled to a delay claim?

Outcome

The owner in this case failed to educate its field personnel of the change in responsibilities that occurred due to DB project delivery. The design-builder had in fact installed the embeds properly, based on shop drawings approved by the DOR. The owner always has the right to stop work if it perceives a problem with quality. However, in this case, that order incurred an obligation to compensate the design-builder for the delay.

Lessons Learned

Large, technically sophisticated public owners like DOTs will normally have an experienced inspection force, as was true in this case. However, the inspection force will often be required to supervise more than one project and, as was the case here, the standards for the project being delivered using DB were contractually different from the ones being delivered using DBB. Thus, it is important to train the owner's field personnel to recognize the differences and be able to adjust their behavior according to the delivery method of the project they are inspecting. Some would argue that owners do not conduct QC inspections on DB projects and that their appropriate role is to conduct quality oversight only. Regardless of the actual quality management policies that are in effect during a given DB project, the owner's representatives must thoroughly understand their roles and responsibilities so they do not inappropriately interfere with the design-builder's construction progress, as happened in this case.

Case 9-3: "If You Have Never Seen It, You Will Never Want It."

The DB RFP for a new military base headquarters building stated: "The final interior design submittal shall include a presentation to the Base Command Group. This presentation shall include but not be limited to: . . . color boards displaying at least two alternatives of actual samples of all interior design treatments and their locations. . . ." The owner explicitly reserved the right to make final approval of all interior design submittals based on the materials shown during the presentation. The base was assigned a new commander shortly after award of the contract, and the new commander had requested that the interior design presentation be made based on the early start date from the design-builder's approved schedule, which had this event occurring during in the later phases of construction. The schedule activity "Prepare and Deliver Interior Design Presentation" had more than three months' float on the schedule that was approved at the beginning of the project.

The DOR's interior design subconsultant indicated that the presentation would be ready as requested. However, the design-builder did not have sufficient time to verify the costs and availability of the various alternatives in the package. The design-builder requested that the presentation be postponed for 30 days to allow time for the various subcontractors to price the alternatives in the interior design package. The new commander insisted on going ahead based on the approved schedule, and the presentation was made as directed.

After the presentation and selection of final interior treatments, the painting subcontractor determined that the selected vinyl wall treatment cost more than the price contemplated in its quote to the design-builder. This material was to cover 75% of the finished wall space. As a result, the sub submitted an alternate material that looked similar and was within the budget. The owner rejected this material. The design-builder, in an effort to satisfy the owner, decided to absorb the additional cost and pay the subcontractor to furnish the original material as specified. When the sub checked on its availability, it found that the required amount was not locally available and to secure sufficient quantity would take six weeks. The construction activity "Install Vinyl Wall Coverings" had only two weeks of float at the time this discovery was made, because changes in the original schedule necessitated by the actual order and delivery dates for other pieces of the approved interior design package had used up about two and one-half months of the original float in this path. The alternate material was immediately available in sufficient quantities. The owner's interior designer stated that if the alternate was accepted, the entire interior design scheme would need to be changed. The design-builder then stated that because the owner knowingly forced it to make the presentation before adequate cost analysis could be completed, the owner should compensate it for additional materials cost and grant a two-week time extension.

The issues that must be resolved in this case were:

- Did the early start date for an activity with float create an obligation by the DB contractor to begin work on that date?
- Was the owner within its rights to require the presentation be made on the early start date?
- Did the design-builder have an obligation to furnish the originally approved material without a change order?

Outcome

An "early start date" on a critical path schedule merely indicates the earliest theoretical date that an activity can start. If it has float, it can also be reasonably started as late, as the schedule's "late start date," without delaying the project. By approving the schedule, the owner merely accepted that the contractor would start these types of activities at some point between the two dates and would not be considered as "behind schedule" and thus liable for any increased retainage or other remedy. Additionally, when an owner reserves the right to make final approval on specific samples, the design-builder must include some factor in its price to account for the selection of higher-priced alternatives. This is normally done by including a contingency in the price proposal and allocating it as required during the project.

In this case, because the chain of events that led to the dispute were initiated by the owner's insistence on an early presentation and decision, the arbitrator found that the contractor was denied the time to react to an unexpected change in the actual price of the wall covering. The decision was to allow the design-builder the requisite time extension with extended overhead, but that the wall covering would be installed at no additional cost to the owner because the design-builder had offered that material as a possible alternative during the presentation.

Lessons Learned

There are several important lessons to be taken from this case. First, the owner was being unreasonable when it insisted on the presentation after the design-builder indicated that it needed more time. What that did was force the design-builder to make design decisions without the opportunity to evaluate their impact. By structuring the contract in the manner described above, the owner was actually asking the design-builder to only present design alternatives that fit its budget and were available in time to support the approved schedule. This is a good technique for sorting out architectural details such as interior finishes. This way, the owner is not offered an unlimited selection of possible finishes but, rather, a selection that the design-builder knows will comply with the requirements of the contract technically, financially, and in a timely manner. Therefore, owner should caution its personnel to not get impatient during DB project execution, and to allow the design-builder to control the schedule. If the presentation had not been made as described, there probably would have been no dispute.

On the other hand, the design-builder gave up its right to additional compensation by agreeing to make the presentation before it had evaluated the alternatives. By doing so, it was essentially stating that the alternatives shown could and would be incorporated into the project without a change to the contract. It could have refused, taken the time to check on the cost and availability of the proposed alternatives, and then made the presentation with alternatives that would have had no negative impact on the project's budget or schedule. Therefore, design-builders must temper their desire to keep an important client happy during a high-visibility DB project, acknowledging the cold reality that their desire to appease the owner might cost them in the end.

The final lesson from this case is quite interesting. One can see that the design-builder initially acknowledged its responsibility to furnish the approved vinyl wall covering when it agreed to compensate the subcontractor for the additional cost of the approved material. This only became an issue when the design-builder found out that to do so would cause the project to finish late and expose it to liquidated damages. This cost was more than it was willing to absorb, so it changed its position and initiated the dispute. The design-builder's initial motivation was influenced by the fact that there were several more attractive DB projects scheduled to be built on this particular military base, so it did not want to "poison the well" with this particular owner. However, when it found that it could not finish the project on time—which presumably would negatively affect its competitiveness on future projects—it changed its stance and let the situation play out as it did.

Case 9-4: "Were You Really Sure You Were Ready for This Project?"

Due to limited and overextended design capabilities within an owner's organization and an owner-mandated emergency requiring rapid delivery of a facility, DB was the selected project delivery system. In the process, the owner used one element of its organization to develop the requirements, outsourced its contracting to another agency, had a third agency provide oversight during the design phase, and had a fourth agency provide oversight during the construction phase.

On this particular project, the overall intent was to place an airport back into operational status. This airport was a secure facility with access restrictions. This project required repairs to the communication systems, rehabilitation of the navigational aids, repairs to the facility's water treatment system, and installation of a new radar system. The communication system was incomplete and required replacement and additional equipment. This updated system was required to address the communications issues between the air traffic control tower and aircraft, the air traffic control tower and the regional Aircraft Control Center (ACC) voice and data, and emergency notifications to the airport crash fire rescue services.

The original contract called for the supply and installation of a VHF and UHF radio system, a voice control and communications switch (VCCS) system, and auxiliary systems. With respect to the navigational aids, the original contract required the supply and installation of technology to meet the certification requirements for an International Civil Aviation Organization (ICAO) CAT II airport. The work included instrument landing system/distance measuring equipment (ILS/DME), an automated weather observation system (AWOS), and an automated terminal information service (ATIS). The work on the water treatment system focused on upgrades to the treatment plant, specifically the chemical dosing equipment, automatic controls, and filter and pumping systems; water distribution system testing and disinfection; one year's supply of chemicals for water treatment; and lab testing and new reverse osmosis (RO) filters. An approach radar and display (ASR-11 and Autotrac) were required to provide safe and near-all-weather landing and departure capacities. The requirement was to meet the latest standards of the ICAO. As construction progressed, additional requirements and scope modifications became necessary to ensure that the system met all requirements of the ICAO.

The issues that had to be resolved in this case were:

- What were the actual owner requirements and when should they have been identified?
- Was the owner ready for this project?
- Was this the right delivery method?

Outcome

Avoidable and unavoidable delays led to cost increases exceeding 40% of the original contract amount. Although the design-builder was nearing the completion of its required scope of work, many issues were yet to be resolved, including (1) training of qualified personnel to operate the systems; (2) provision of reliable electric power; (3) recertification of systems in accordance with ICAO standards; and (4) long-term operation and maintenance.

The aforementioned items resulted in the following outcomes. First, the appropriate end-user personnel did not receive the hands-on training by the equipment and software manufacturers. Second, the navigational aid flight check was successfully performed but a recent loss of electric power at the site presented a setback in the fact that the original flight check was now void and had to be performed again prior to final certification by ICAO. This resulted in additional costs but there was not sufficient funding available to pay for this recertification. Third, the local authorities were not able to address the water problems in systems that connected to the airport facilities. This made the installation of the water treatment facility ineffective, so the water treatment plant was removed from the project. A new modification was planned to eliminate the original water treatment plant repairs and upgrade and to provide a new package plant water treatment system, which would treat water for the chillers. The projected savings from the water treatment plant components were de-obligated from the contract and re-obligated to fund cost growth on the radar installation. The radar was scheduled for a check by the U.S. Federal Aviation Administration (FAA).

Initial project progress was good. Design and equipment submittals progressed well and the design-builder demonstrated good technical capability to design and execute the contract. However, time delays quickly came into play for the following reasons. First, major security issues in the vicinity of the airport stopped all operations and forced demobilization of design-builder personnel for a period of approximately two months. This added to other delays and project issues.

Second, the highly specialized equipment necessary to accomplish the work had very long lead times for acquisition and delivery to the site. This created project delays when the necessary equipment could not be received at the project site in the original anticipated time frame.

Third, the extended length of the contract caused significant cost increases to the project. This forced the owner to obtain additional funding through proper sources. The acquisition of those funds proved time-consuming and forced the partial demobilization of design-builder personnel. Fourth, due to the specialized equipment, training was required for end-user personnel. This presented serious difficulties in working through the various client entities and the local airport authority. Training was held in three different countries, at the equipment manufacturers' sites. Obtaining visas for the local personnel who were going to operate the systems proved to be very difficult and time-consuming. Training sessions were cancelled as many as three times due to inability to obtain travel documents. Unless those issues were fully resolved, the project had the potential to be completed without trained personnel to operate all the systems.

Fourth, early in the project, a lack of reliable electric power was noted and concerns were passed to the local entity responsible for overall airport management. A separate power contract was initiated, but conflicts arose when the power contract could not deliver a reliable source of electric power for the various systems in a timely manner.

Fifth, there was lack of support from qualified owner personnel. The extended contract time and the cost increases also led to budget restraints on the part of the contracting entity. As a result, the owner's QA inspector was removed from the job a year before completion. Lack of oversight by qualified personnel left the design-builder without adequate interface with and assistance from the owner's personnel. Several delays in the project were attributed to lack of timely submittals by the contracting officials to the correct entities within the owner's organization.

Lessons Learned

The primary lesson to be learned with this case study is how important it is to adequately define the owner's requirements and to consider the risks resulting from outside influences. To prevent these types of problems in future DB cost-plus contracts awarded in a contingency environment, several factors should be considered prior to award and included in the contract documents. First, contract time—especially in a cost-plus contract—directly contributes to the design-builder's ability to perform its scope of work within the initial budget. Any delays, whether avoidable or unavoidable, will carry an associated cost increase and affect the final completion of the project.

Other issues to be considered prior to contract award are:

- Identify training needs for the end user, and identify and rectify any logistical or political roadblocks that might interfere with that training. Training should take place early in the project to avoid any unforeseen difficulties at contract end.
- Identify and investigate the suitability of all facility services (electric, water, sewer, etc.) that will be required for support of the project. Prepare contingency plans should these services not be available.
- Identify and prepare for site-specific factors that would affect production (security, permits, etc.). These should be prepared for in the contract period of performance and the design-builder made aware of their potential impact.

- Long lead items should be identified in the planning stage and specifically addressed in the contract documents for review, approval, and acquisition.
- Regular meetings between the owner, engineer, design-builder, and end user are critical to ensure that the design-builder is meeting the initial schedule. Any potential delays should be identified and mitigated as soon as possible to ensure that the contract schedule stays on track, reducing potential cost increases.
- Plan and prepare for long-term operation and maintenance of critical components define in the contract.

Case 9-5: "Whom Are You Trying to Satisfy?"

This DB project in a foreign country was intended to restore selected locomotive, freight wagon, and passenger coach maintenance facilities to useful operation and good repair, including the main locomotive shed, the passenger coach technical station, associated outbuildings, and other selected facilities. The work consisted of all required electrical, plumbing, and civil elements needed for repair and specialized railway inspection, repair tools, cranes, machines, and facilities.

To ensure that the work was completed in a timely manner and that all interested contractors were aware of the requirements, two mandatory site visits were conducted prior to award. Attending parties included representatives from the financing organization, the owner's personnel, and representatives from several construction firms. The contract was awarded to a small business that was expanding its DB capabilities. Construction started upon approval of the preliminary design by the financing entity and the owner.

Concerns about the project arose almost immediately. Societal prejudices immediately affected the design-builder. The owner of the company was a woman who was from a region of the country that was not considered "acceptable" by the personnel who operated the rail facilities, even though the financing entity and the contracting entity considered her and her firm to be competent and the best firm for the project. The design-builder was at a disadvantage in a male-dominated society in an area with a strong dislike for people from her region.

A common issue for many projects in strife-torn areas is disagreement concerning who truly represents the owner. In this particular instance, the personnel who operated the rail facilities believed they were the true owner representatives, while the financing organization and representatives of the national government believed *they* represented the owner. Unfortunately, the contract language gave credence to each entity's claim that they were the voice of the owner.

Although certain design-builder personnel activities were part of the design and construction approval process, the personnel at the rail facilities refused the design-builder personnel access to the site on several occasions and for extended periods. Wanting full control of the project, the facility personnel struggled with the design-builder and the other members of the owner's team. The contracting personnel sent a letter to the facility personnel requesting a stop to the delays, which at that point totaled more than 212 days and more than $350,000 in additional project costs. These delays extended the contract required completion date by several months. Delays induced by the railroad personnel continued. With the contracting agency unable to resolve the control issues, they eventually terminated the contract for convenience.

Lack of the original design drawings for the facility before renovation also caused problems on this project. Can a contractor be expected to easily bring a facility back to its original design and status in an expedited manner without the original as-built drawings? For this project, project bidders were informed that the railroad would provide pre-renovation as-built drawings at the bidder's request; however, they were warned that the accuracy of any such representations was potentially flawed. What further confounded the situation was the simple fact that the contract required design standards that conflicted with aesthetic and renovation requirements. Trying to resolve these issues among the various "owner" entities proved impossible.

Many small contractors are not familiar with the regulations and procedures inherent in DB contracts. This issue is exacerbated when the contracts are written in a language different from the language used for most communications in the project area. In this case, unfamiliarity with required contracting procedures resulted in inadequate submittals, delays in approval of work, and delays in contracting actions such as claims and modifications. The contracting authority had concerns from the outset about whether the design-builder understood what was required, especially in terms of the proposal requirements and the subsequent submittals.

Several delays in this project were caused by police curfews and incidents resulting from civil unrest which restricted access to the project site. Also, skilled labor was scarce due to lack of security and the reality that many experienced engineers had fled the area. This caused quality concerns at this project and many others. Finally, there was not a great comfort level with the International Building Code provisions. Local contractors paid lip service to the requirements and typically fell back on outdated and no-longer-accepted methods, which often resulted in a rejection of substandard work.

The issues that had to be resolved in this case were:

- Who spoke for the owner?
- Were the appropriate requirements identified in the contracting strategy and in the selection process?

Outcome

From almost the onset of the project, the design-builder was seriously behind schedule. Numerous attempts were made by the design-builder to work out the issues with site personnel. The design-builder was also proactive in trying to work with the contracting and financial entities to resolve site and contract issues. The design-builder was unable to maintain the required standards of quality because the site personnel refused to allow it to bring in the experienced members of her team. Combined with the increasing demands and continued obstacles created by the site personnel, both the design-builder and the contracting entity decided it was best to terminate the contract for convenience.

Lessons Learned

First, the owner's team needs to be a *team*. The contract needs to effectively spell out the roles of the owner regarding with whom the design-builder will communicate and coordinate. The end user and the entire owner team should be closely involved with the original development of project requirements and in the design review process. Inclusion of owner representatives will prevent unnecessary modifications and frustrations while construction is underway.

Regular periodic meetings should be held between the end user, a proactive financial entity that wants a voice in the owner's organization, and the design-builder. In this case, this would have allowed concerns to be aired and addressed *prior* to when the end user blocked access to the site and caused serious unscheduled delays and cost increases.

Second, the contract language and the corresponding requirements need to acknowledge social realities. Societal prejudices may need to be addressed prior to any contract award due to regional discriminatory attitudes. DB is not a mechanism to alter social prejudices.

APPENDIX A

Implementing Building Information Modeling in Design-Build Projects

Tamera L. McCuen

As the architecture/engineering/construction (AEC) industry advances into the twenty-first century, facility owners are seeking methods for new construction of capital assets in which the design, construction, and operations of facilities are more efficient and eliminate unnecessary waste of both material and human resources. Therefore, owners are increasingly choosing design-build (DB) for their project delivery method and building information modeling (BIM) as the tool that maximizes the benefits inherent in the DB method. DB is a dynamic, interdisciplinary delivery method in which the communication and decision-making processes are executed in an expeditious manner which recognizes that time is money. BIM is a dynamic, interdisciplinary method for information exchange facilitated by computer technology. Owners requiring BIM for new construction projects include the U.S. Army Corps of Engineers (USACE), the U.S. General Services Administration (GSA), the U.S. Department ofVeterans Affairs (VA), and the National Institutes of Health (NIH).

In a 2007 survey, more than 200 North American and international owners indicated both an increase in the use of DB project delivery and the use of BIM for design, construction, and facility management (FMI/CMAA 2007). More than 17% of those responding to the survey utilized DB in a majority of projects, and approximately 35% of those responding indicated that they utilize BIM due to its benefits of improved communication and collaboration.

BIM is a system in which improvements to traditional business processes combine with a system of computer technology to facilitate a collaborative environment from conception of a project throughout the facility's life cycle. BIM provides a venue that brings the entire team consisting of the owner, designers, constructors, and the facility manager together from the project onset for efficient and effective

Tamera L. McCuen, LEED AP, is Assistant Professor of Construction Science at the University of Oklahoma.

decision making. DB project delivery is the ideal method for projects using BIM because both DB and BIM emphasize communication and collaboration.

Defining Building Information Modeling

There are as many definitions of BIM as there are professional associations within the AEC industry. However, the National Institute of Building Sciences (NIBS)'s national BIM standard provides a definition of BIM that is useful because it does not focus on one user group's perspective or working application of BIM; rather, it is a global definition. This definition clearly states that BIM is more than just software; it is a *process* inclusive of all stakeholders and based on collaboration:

> Building Information Modeling (BIM) is a digital representation of physical and functional characteristics of a facility. A BIM is a shared knowledge resource for information about a facility forming a reliable basis for decisions during its life-cycle, defined as existing from earliest conception to demolition.
>
> A basic premise of BIM is collaboration by different stakeholders at different phases of the life cycle of a facility to insert, extract, update or modify information in the BIM to support and reflect the roles of that stakeholder. (NIBS 2007)

One of the main drivers leading the demand for improved efficiency in the AEC process, and thus supporting DB and BIM, was an eye-opening 2004 Bureau of Labor and Statistics report. That report revealed an actual decline in the productivity rate in the construction industry since 1964. Owners and leaders in the industry found this report unacceptable, especially given that the decline in that rate was in direct contrast to the "other non-farm productivity rate index," which indicated productivity improvement at record rates. Consequently, the owners responded by demanding more efficient processes and are now increasingly choosing DB. In addition, several owners are requiring the use of BIM on all new construction projects. This trend toward BIM leaves facility owners and AEC professionals wondering where to start. The first steps begin with assembling the project team, selecting the appropriate software, and prescribing a method for BIM implementation in DB project delivery.

The Team

Inherent in the DB method is a focus on efficiency, which is accomplished by bringing together all the team members from the onset of a project for a unique interdisciplinary approach. Assembling the DB project team may occur either in-house or between firms, but in any case gathers multi-functional tasks into a decision-making process focused on efficiency. Borrowing from DeSanctis's and Poole's idea of a technology's "spirit" independent of its structural features will help form a clearer understanding of the parallels between DB and BIM. A team that embodies the collaborative spirit of DB from the beginning is one that is more likely to benefit from the spirit of BIM as a tool to utilize all the stakeholders' input throughout the process (DeSanctis and Poole 1994). What follows is a brief discussion about the spirit of BIM, which will provide a better understanding of how it relates to the DB team.

The spirit of a technology refers to the decision process, leadership, efficiency, conflict management, and overall atmosphere of the technology. Each of these characteristics is a benefit of using BIM as a tool for improved project delivery. "Spirit" is the general intent of the technology with regard to the values and goals underlying the technology. The underlying values and goals of BIM are to ensure accurate and efficient information exchange between stakeholders and across disciplines. In the past, disciplines were founded on and resided in the functional silos of design, construction, and operations. The emphasis on information exchange makes it clear that the "I" in BIM is important. Traditionally, accurate and efficient information has been the missing piece between stakeholders and, consequently, has led to inefficiencies plaguing the AEC industry. Ultimately, DB benefits from BIM in the convergence of the interdisciplinary DB team focused on efficiency, supported by an interdisciplinary tool focused on efficiency.

The Technology

BIM is a computer technology based on a system of interdisciplinary users, task-specific software applications, and computer hardware. BIM is a digital representation of the 3D, 4D, and 5D information about a project. Within the 3D aspects of a BIM is the graphical representation of a facility with its element properties defined and attached to each assembly and component. 4D refers to the time and the schedule of a project. 5D is the quantification process and subsequent cost information about a project. The previous section discussed the interdisciplinary team within the context of DB and BIM, providing the necessity for assembling a BIM system that is interdisciplinary. This section focuses on task-specific software and considerations needed prior to selecting one application over another. It is important to clarify that BIM is not a software program but, rather, a process and a way of collaborating and communicating that requires a system of software.

Information contained within a BIM varies from project to project, thus making it difficult to provide a precise checklist for the exact information output one can expect from a model. As a tool, BIM is designed to meet the owner's program for a project as well as support the facility's operations throughout its life cycle. Given these parameters, it becomes clear that each BIM is as unique as the facility itself. For example, one facility may require complex structural analysis for a complicated design, while another facility may require an extensive energy analysis.

Within these scenarios, the first project may require structural design and analysis tools that accommodate complex shapes and undulating surfaces for which the designer would choose to use an application that fulfills these parameters. Another, more traditional linear design would require structural design and analysis that could be accomplished with a different software application. Therefore, choosing BIM technology for design and analysis is contingent on the project itself.

Because each project is unique, and because there are many options for BIM software, it is important to discuss the concepts of interoperability and integration. "Interoperability" is accomplished using Industry Foundation Classes (IFCs) bridging all software applications. "Integration" is limited and proprietary between certain software applications. Because members of DB project teams are chosen based on the program requirements of a project, it becomes imperative that all software needed to fulfill each team member's task be interoperable. Without interoperability, the software will limit the participants on a BIM team to only those who have certain proprietary applications.

The term "interoperability" is often interchanged with "integration"; however, there is a distinct difference. Interoperability is defined as a seamless data exchange between diverse software application types in which each may have its own internal data structure but still allows for accurate data exchange. Interoperability is essential and, although the technology to date is not perfect, software developers working to develop products that are increasingly interoperable. Interoperability allows each discipline to perform its functional tasks knowing that the information with which they are working is reliable and accurate. DB projects using BIM as a tool for decision making is especially reliant on the technology's interoperability, given the interdisciplinary, fast-paced decision-making process. Consequently, the DB delivery method may be especially vulnerable to the lack of interoperability if team decisions are made on unreliable, inaccurate information contained in a BIM whose applications are not truly interoperable. A BIM system may still function simply through integration, but users of such systems are warned that the information contained within it is only as reliable as each discipline's validation of all other discipline inputs prior to using the model's information to perform their tasks. This need to go back and validate exchanged information defeats the efficiency of BIM and the DB process, and only further perpetuates the lack of improvement in the construction industry's productivity rate.

For example, the designer's software application and the cost estimator's software application must be interoperable to ensure the data exchange is accurate to support each discipline's tasks. Within a BIM, the designer's 3D representation of the facility includes information about the physical characteristics of each assembly and component, which the cost estimator can extract to execute the quantification and cost processes for that project. Without interoperability, data may be lost in the transfer or the data may be incorrectly transferred. Both scenarios are precarious and lead to an invalid model. Software *may* be able to integrate the information, but integration does not necessarily provide for interoperability. As such, it can put constraints on the project which could be avoided if the software was interoperable. Constraints from lack of interoperability include (1) limits placed on who can provide information in a BIM; (2) what the project's information output might be; and (3) the accuracy of information exchange in a BIM. The best options in software applications are those developed for interoperability using IFCs.

Computer hardware requirements are directly related to the software applications utilized by each discipline. A model server is needed on which the BIM may reside. The lead entity on a DB project will most likely provide the model server, whether it is a server in-house or a web-based service. BIMs contain a large amount of data and thus require sufficient processing power and memory capacity to facilitate use by each team member.

By this point it should be evident that the success of BIM is contingent on accurate information exchange within the context of the team as well as the technology system assembled. Implementation of BIM into a DB team can be met with resistance by team members rooted in uncertainty and inexperience. However, a strategy for BIM implementation based on collaboration and communication can prove successful. To demonstrate this point, here is an example from a case study of one of the 2007 American Institute of Architects (AIA) Technology in Architecture Practice BIM award winners.

The Implementation

The 2007 recipient of the Design/Delivery Process Innovation Using BIM award was the M. A. Mortenson Company of Seattle, Washington for their Benjamin D. Hall research building. Mortenson used BIM not because it was required by the owner but because it was a design-build-operate-maintain (DBOM) project delivery, and Mortenson could rely on BIM to provide a model containing reliable information from an interdisciplinary team which would support decision making for the facility going forward. Remember that the definition of BIM is a reliable source of information for decision making from conception through the life cycle of a facility.

A case study of this project, supported with interviews of Mortenson's project design coordinator, revealed an implementation process based on robust communication and collaboration. The design coordinator served as the team's leader and BIM coordinator. Team members were hand-picked to ensure that they were the individuals most willing to lead and accept change, with no regard for previous experiences among the team members. Consequently, the assembled team began as a new team with no experience of directly working together. The unofficial motto for the team was "If you always do what you always did, you'll always get what you've always got." This motto was pervasive and it served the team throughout the process. The overarching characteristics of the team were based on collaboration, rapid decision making enabled by using the technology, and individual team members' willingness to accept change. Each of these characteristics reflects the spirit of both the DB delivery method and BIM. This ultimately led to the team's success. According to Mortenson's design coordinator, DB is the "best fit" project delivery for the use of BIM as a tool for collaboration and communication.

In the 2007 FMI/CMAA report, owners identified the top two benefits of BIM as improved communication and collaboration among project participants, and higher-quality project execution and decision making. The top two barriers to BIM ranked by owners were lack of expertise and greater system complexity (FMI/CMA 2007). Benefits of choosing the DB project delivery method mirror the benefits of choosing BIM. The roadmap to success for owners in the twenty-first century includes a combination of the right DB project team, the right BIM technology, and an implementation strategy focused on collaboration and communication.

References

DeSanctis, G., and Poole, M. (1994). "Capturing the complexity in advanced technology use: Adaptive structuration theory." *Organization Sci.* 5(2), 121–147.

FMI/CMAA. (2007). "Eighth annual survey of owners: The perfect storm—construction style." FMI Corporation, <http://cmaanet.org/foundation-research-projects> [April 10, 2008].

National Institute of Building Sciences (NIBS). (2007). "United States National Building Information Modeling Standard, Version 1. Part 1: Overview, principles, and methodologies." *National Institute of Building Sciences, Facility Information Council*, <http://www.wbdg.org/bim/nbims.php> [Dec. 18, 2007].

APPENDIX B

Implementing Sustainability in Design-Build Projects

Lee A. Fithian

Sustainable design and construction features are becoming more common and have become mandatory for many public and private projects. The U.S. Green Building Council (USGBC)'s Leadership in Energy and Environmental Design (LEED) certification is often used by public agencies and private owners as a means to articulate their desire to design and build projects that are both energy-efficient and environmentally responsible. Furthermore, the USGBC and the American Society of Heating, Refrigerating and Air-Conditioning Engineers (ASHRAE) are participating in developing a code-based system that can be adopted by reference by governing authorities.

LEED certification shares the stage with Green Globes (a rating system developed by the Green Building Initiative) as one of the two pre-eminent methods of independent certification of sustainability, but these standards have been considered to be less progressive than alternate methods that take into account carbon neutrality or positive energy. However, the benefit of establishing a sustainability rating system, such as LEED, as a criterion is that it can be used as a metric to evaluate construction options regardless of whether LEED certification is sought for the project.

While third-party evaluation of sustainability is useful, it is also a moving target. The evolution of the standards can cause concern regarding the regionalization aspects during construction. It becomes imperative that the local markets, means, and methods play a larger role during the scoping work of the design-build (DB) entity as it seeks to meet sustainability criteria and certification levels.

Lee A. Fithian, AIA, NCARB, AICP, LEED AP, GaC, is Assistant Professor in the College of Architecture at University of Oklahoma.

Regional or Local Adaptation

Local codes, typically the watershed for establishing baselines for the subcontracting submittal and incorporating them into the bid, can become irrelevant as they are superseded by imposed sustainability criteria. This is true especially with regard to energy efficiencies, material procurement, and construction waste management. The importance of the local "boots on the ground" subcontractor playing a more substantial role in the project scope definition is not necessarily typical practice but must be taken into account.

The form of the DB entity will also have to take into account these local subcontractors who will be not only performing the work, but also providing the documentation necessary to achieve the sustainability certification. Although the LEED standard has become prevalent in larger markets, as the DB project delivery and integrated practice delivery methods standardize in the industry, teaming requests based upon local and regional availability must be couched in terms of understanding the additional requirements.

Material Selection and Construction Waste Management

Regional adaptation awareness is also imperative to inform the construction process; it must be planned for early in the project delivery. Material selection, particularly with regard to construction waste management, is fundamental in reducing waste. The documentation requirements for material procurement and waste management are formidable, yet potentially profitable, in yielding installation efficiencies and reducing shipping costs.

The prevalence of construction waste management as an owner criterion has led to the establishment of specifications regarding the allocation of construction waste costs savings—sometimes to the owner and not the DB entity. Understanding this owner criterion as well as properly estimating material quantities can lead to "green" procurement opportunities. Further incentivizing construction waste management has led to state-level establishment of sustainability criteria. The state of Massachusetts, for example, has recently banned construction waste in its landfills.

Construction waste management can also affect construction means and methods. Identifying local recycling opportunities can lead to on-site adaptations. For example, the lack of localized recycling can create the opportunity for centralized cutting areas that yield material efficiencies. Labeling recycling bins from above or on the side aids in construction phasing, and investigating on-site sorting operations for dense sites could be explored. Material diversion for landfill includes documentation organized by weight, with conversions for volume, and methods should be developed accordingly.

The DB entity, particularly with regard to overseeing those aspects of the work that are not self-performed, must be extremely vigilant regarding the tendency to substitute materials. Green specifications calling for the use of specific materials with specific chemical compositions and installation methods contribute to sustainability certification through documentation and commissioning (refer to the Commissioning section below). Material safety data sheets (MSDSs), cut sheets, and bills of lading are typically required to document the material contribution, but the commissioning phase may call for pressurization tests and air sampling. Materials such as paints, adhesives, and sealants can be particularly problematic in this regard.

While material substitution can be problematic, the impact on local operations and businesses can be significant. Regionalized procurement criteria can also overtax localized resources. Stimulating local business opportunities is a positive by-product; for example, the recycling of gypsum board is fairly localized yet the opportunity for its use as an admixture for agricultural waste is typically ignored. Understanding that businesses can work with materials produced outside the region yet fabricated locally can also be an important factor. Recognition by the constructor that local and regional materials also have lower transportation costs (with inherent lower embodied energy and reduction of corresponding pollution) is fundamental in managing overall costs during project delivery.

Along with these components of construction means and methods, cement admixtures and concrete formwork and placement that are self-performed must be addressed. Fly ash and recycled paving need to be characterized, incorporated, and their use documented because they can provide significant contributions to the recycled content of the project.

Construction Indoor Air Quality

Construction operations that incorporate sustainability criteria also address indoor air quality concerns, both during construction and after substantial completion. Dust and fumes from work activities should be taken into account because their effects are mitigated off-site. Construction workers and their environment are addressed by sustainability criteria. The proper sealing of mechanical systems and change-out of filtration media, both during and after construction, will need to have proper documentation, some of it photographic.

The development of a construction indoor air quality plan is also part of most sustainability criteria and provides documentation for the commissioning phase. Maintaining positive air pressure in areas not under construction or renovation can lead to construction sequencing changes. Building flush-outs, if provided for in the green specifications, must not only be planned for from a delivery time aspect, but will also involve filtration media and cessation of most typical (carpet and furniture installation) close-out procedures. Indoor air quality plans can also require isolation areas for materials to prevent point source pollution (e.g., adhesives and plumbing sealants).

Material procurement, on-site material protection, installation, and cleanup all play roles in maintaining indoor air quality. Understanding the roles of various manufacturing processes and materials can assist in the procurement and substitution of generic materials brought onto the job. Indoor air quality problems stem from urea formaldehyde in engineered wood products; volatile organic compounds (VOCs) in paints, sealants, and adhesives; dust, animal dander, and pollens; radon and carbon monoxide levels and high indoor humidity levels; and many cleaning products. Before project "dry-in," condensate from mechanical systems can drip in high-humidity situations, which can introduce mold and mildew prior to substantial completion. Vinyl wall coverings with so-called green glues that do not inhibit mold growth can mask mold and mildew. Improper protection of stored materials can introduce pollutants, and installing drywall before closing in the building can exacerbate moisture problems.

Sites, Public Projects, Alternative Technologies, and Acquisitions

The DB project delivery method can result in an inherent coordination of design and performance with potential for accelerated economic returns for sustainable systems performance. The public entity owner can clearly articulate its desires regarding sustainability by assigning weight in relation to other factors in the DB evaluation plan, particularly with regard to a two-phase selection process.

Site work and mobilization will be altered by sustainability criteria. Reducing pollution and limiting erosion of topsoil are typically the outcomes addressed. Stockpiling topsoil, identifying staging locations, and preventing runoff and sediment from entering overtaxed stormwater and naturalized waterways through the development of a comprehensive stormwater management plan are typical criteria. They may or may not be part of a typical DB entity's service offerings, but the plan documentation and demonstration of implementation are part of the submittals necessary to achieve sustainability certification. A locally approved stormwater management plan may not be adequate for sustainability criteria since it must exceed local zoning requirements. Altering construction sequencing to complete site work early presents opportunities to prevent erosion and runoff and provide for subsequent staging operations.

Some projects with sustainability criteria may require the use of alternative technologies and systems. The DB entity should familiarize itself with these types of systems, seeking historical data and actual installation performance prior to submitting pricing. Furthermore, operations and maintenance (O&M) personnel may be unfamiliar with alternative systems requirements. For example, some materials may require alternate maintenance procedures, and some systems controls may incorporate technologies requiring specialized training that may be beyond the scope of the initial proposal.

Coordination and sequencing of subcontractors can be altered by newer technologies and alternative systems. For example, lighting systems can be tied to networking and communications devices for remote operations and automatic dimming based on daylighting solutions. Contravention of these technologies can hamper energy efficiencies, considering that lighting also produces heat which taxes the cooling system.

Regulatory controls are not universal, especially with regard to wastewater treatment and reuse. Opportunities that may have been incorporated during the design phase may not meet approval during construction, particularly with regard to low-impact site development and rainwater collection for toilet flushing. Local permitting may be delayed, causing construction delays should officials be unfamiliar with alternative systems that are acceptable in many other regions.

Particular attention should be paid to the selection of roofing and plumbing alternative solutions. Operational efficiency of potable water usage is typically tracked as part of the sustainability criteria. Attempts to prevent call-backs by substituting standard urinals, fixtures, and faucets can increase potable water usage dramatically and prevent meeting sustainability criteria. Systems thinking must be inherent in the materials substitution process; for example, stormwater runoff calculations can be affected by altering green roof material selections.

The design schedule can also be affected by public participation processes. Bypassing these can limit social equity issues that are inherent in many sustainability criteria systems. Due to the normally time-consuming processes associated with state, municipal, and federal requirements for mandatory announcement and the convening of public hearings, certain sustainability measures such as wetlands mitigation and avoidance of undeveloped areas raises concerns for eminent domain and brownfields redevelopment, which can affect construction contract performance time.

Commissioning

Sustainability criteria imposes a mandatory commissioning phase, sometimes termed the "sixth phase" which, while normally part of public projects, is not normally a requirement in private or state- and municipal-level sectors. ASHRAE defines commissioning as:

> . . . a quality-oriented process for achieving, verifying, and documenting that the performance of facilities, systems, and assemblies meets defined objectives and criteria. Commissioning is an all inclusive process for all the planning, delivery, verification, and managing risks to critical functions performed in, or by, facilities. Commissioning ensures building quality using peer review and in-field or on-site verification. Commissioning also accomplishes higher energy efficiency, environmental health, and occupant safety and improves indoor air quality by making sure the building components are working correctly and that the plans are implemented with the greatest efficiency. (ASHRAE 2005)

The availability of local subcontractors to the commissioning authority, which is typically an entity independent of the DB, can cause confusion and potential cost increases if these criteria are not explicitly lined-out during the negotiation phase. Those DB entities that do not self-perform the mechanical and building controls aspects of the project delivery will be well served to determine which localized or regional groups are best suited to perform such services for each project, since local climate issues play an all-important role in determining which mechanical systems solutions yield the greatest efficiencies. Furthermore, these local or regional entities must be brought on early enough to participate in establishing project close-out procedures and the O&M documentation necessary to fulfill the fundamental commissioning requirements.

Coordinating multiple trades is a major strategy in improving the performance of the building envelope. Unfortunately, performance of the building is not tested and verified until the commissioning phase. Each trade must understand its contribution to the overall sustainability rating of the exterior closure. Construction inspection needs to be ramped-up to validate proper sealing of penetrations and connections, and elimination of opportunities for thermal bridging. Project energy performance and pressurization tests will point out failures in these categories, and the cost of call-backs to correct them can be prohibitive.

The highest potential for significant harm to the DB entity during the commissioning phase is a commissioning report provided by a certified commissioning authority that is unfamiliar with the means and methods—particularly alternative systems and technologies—built into the project. Great care should be given to early selection of and coordination with the commissioning authority so that all design and performance issues are understood, and that installation methodologies are tracked and documented by the commissioning authority to prevent misunderstandings.

Design-Build with Sustainability not Part of the Contract but Voluntarily Incorporated by the Constructor

The opportunity for construction procedures to incorporate sustainability criteria without owner-provided requirements are numerous. Reference to the previously discussed materials selection opportunities can yield significant cost savings. Implementing construction waste management practices can actually yield profits from diverting landfill costs, materials procurement efficiencies, and recycling the provider's purchase of waste material.

Depending on local conditions, site infrastructure could be reduced and cost savings found through the application of alternative paving systems. Porous and permeable paving systems do not use traditional curbs, and bio-swales and rain gardens contribute to infiltration that reduces the sizing of stormwater systems. Furthermore, rainfall catchment and landscape material selection, also inherently local, can reduce and/or eliminate the need to install irrigation systems.

Fluctuating costs associated with the oil industry can lead to the use of concrete-based paving systems that can also affect project cooling system performance due to reduction of thermal island effect. An understanding of the local economy, inherently in the purview of the local subcontractor, can contribute to the project pricing strategies during the design development stages. In the case of a DB entity that will necessarily own and operate the facility, tighter integration with the entire DB process delivery method can deliver significant operational cost savings.

Design-Build with Sustainability Requirements Applied after Contract Award

It is highly problematic when sustainability requirements are applied after contract award. The integrated systems approach for the development of a sustainable design solution can be thrown off schedule due to the imposition of criteria not normally accounted for in the DB project delivery method. It is imperative that the DB entity identify and, if necessary, educate the selection board or owner regarding code adoption or legislatively imposed sustainability criteria.

Project owners and public project selection entities are not necessarily familiar with regionally responsive strategies and can call for criteria which, while highly successful in other regions, are inefficient or untenable in the project development area. The DB entity should be prepared to review the owner criteria and suggest alternatives that can achieve sustainability rating levels even though they may not match specific criteria called for in the RFP. Furthermore, it would be advantageous for the DB entity, if at all possible, to participate in and/or track public policy developments in all regions and local areas where work is to be performed in order to help shape the scope of sustainability criteria and their adoption.

References

American Society of Heating, Refrigerating, and Air-Conditioning Engineers (ASHRAE). (2005). *Guideline 0-2005: The commissioning process,* ASHRAE, Atlanta, Ga.

Index

administration, construction 255–266
administration, design 65–95; charettes 81–85; culture change 66–71; design document quality 69–71; design personnel 71–75; design process review meetings 91–92; design reviews 92–93; design submittals 89–93; drawing numbering system 93–95; integrating charette design input 85; integrating contract design input 75–81; integrating design constructor input 77–81; integrating design subconsultant products 76–77; internal design process management 67–69; overview 65–66; performance criteria interpretation 85–89; responsibility chart 72–74; trade subcontractor design input 81; value analysis 80–81; value engineering 80–81
administration, project 14–18
BIM. *see* building information modeling
building information modeling 267–272; defined 268; implementation 271–272; team members 268–269; technology 269–271
building sector 3
case studies 149–161; articulating technical preferences 255–257; changing contractual scope of work 157–159; construction administration 255–266; defining owner's requirements 261–264; non-licensed design professional 154–157; resolving differences of opinion 159–161; roles and responsibilities of the designer-of-record 149–152; scheduling issues 258–261; social issues 264–266; training field personnel 257–258; transferring design liability 152–154
change orders 191–217; best practices 198–200; change management 196–200; during construction 206–208; before construction starts 201–203; after construction starts 204–206; costs of 211–215; during design 200–217; design errors and omissions 209–211; direct costs 212–213; entitlement 194–196; impact costs 213–215; indirect costs 213; overview 191–193; schedule-related costs 213; scope changes 208; timeline 196–198; types of 194–195; unforeseen conditions 208–209; value engineering changes 215–217
charettes 81–85; defined 81–82; integrating charette design input 85; preparation 82–85
communications protocol 32–33
constructability 15–16; input to the design 77–80
construction administration. *see* administration, construction
construction quality. *see* quality, construction
contracts: analyzing requirements 39–42; coordinating scheduling with 55–56; models 9–12; scheduling clauses in 40–41

control: cost 248–252; document 17–18; horizontal 177–178; project 16–17; vertical 177–178
cost control 248–252
DBIA 2
design administration. *see* administration, design
design liability 120–123, 152–154
design progress payments. *see* payments, design progress
design quality. *see* quality, design
design review 138–143; checkpoints 136–138; philosophy 13–14; resolving differences of opinions during 126–128; responsibility for 143–146
design-bid-build projects: constructability 15–16; contract models 9–12; design review philosophy 13–14; document control 17–18; performance criteria 12–13; project control 16–17; safety oversight 18; trust in 18–20
Design-Build for the Public Sector 4
Design-Build Institute of America 2
design-build projects: background 1–4; compared to design-bid-build 9–20; constructability 15–16; contract models 9–12; design review philosophy 13–14; document control 17–18; factors for increased use 5–6; performance criteria 12–13; preeminent factor 27–29; project control 16–17; research 4–8; safety oversight 18; success factors 33; trust in 18–20
design-build team of the owner 21–24
design-builder: perspective on design submittals 89–90; perspective on retainage 231–232; project team 24–26
designer-of-record 71; activity definition 46–52; activity scheduling 51–52; promulgating performance criteria 87–89; roles and responsibilities of 149–152
document control 17–18
drawing numbering system 93–95
modeling, building information. *see* building information modeling
owner: activity scheduling 44–46; assumption of design liability 120–123; defining requirements 261–264; design quality role 107–108; design-build team 21–24; perspective on design submittals 90–91; role in defining construction quality requirements 168–172
partnering 29–33; communications protocol 32–33
payments, design progress 221–230; design commitment 222; determining progress 226–229; payment for construction progress 226; payment for materials 229–230; payment formula for design completion 222–226. *see also* payments, progress
payments, incentive 240–248; access incentives 243–244; award fees 242; environmental impact reduction incentives 244–248; quality incentives 242–243; schedule incentives 241–242; traffic incentives 243–244
payments, progress 219–252; cost control 248–252; for design progress 221–230; final payment 230–235; guarantee maximum price 237–239; incentive payments 240–248; limitation on payments 234–240; overview 219–221; payment for materials 229–230; payment formula for design completion 222–226; priced cash flow 240; retainage 230–235. *see also* payments, design progress
performance criteria 12–13; criteria dictionary 86–87; interpretation 85–89; promulgating by designer-of-record 87–89
PQA. *see* project quality assurance model
procurement laws 2
progress payments. *see* payments, progress
project administration. *see* administration, project
project control 16–17
project quality assurance model 101–102; application of 104–106
project teams 20–27; design-builder's team 24–26; owner's design-build team 21–24; summary 26–27
quality, construction 163–188; allocating quality assurance responsibilities 173; in the construction phase 172–175; construction submittal review 176–177; defined 166–168; design-build issues 173–175; economics of quality 172–175; horizontal control 177–178; management planning 176–183;

nonconforming work 181–183; overview 163–188; owner's role in defining requirements 168–172; performance quality objectives 169–172; personnel requirements 186–188; post-award construction quality plans 185–186; quality control testing 180–181; resolving differences of opinions 173–174; routine construction inspection 178–179; scope creep 174–175; types of 166–168; variances from approved submittals 175; verification and acceptance testing 183–185; vertical control 177–178; witness and hold points 179–180

quality, design: attitude change 118–120; changes to submitted designs 131–132; defined 106–118; design review checkpoints 136–138; design reviews 138–146; design submittal process 136; design-build issues 120–146; economics of 118–146; general requirements 134–136; management planning 132–133; objectives 108–118; owner assumption of design liability 120–123; owner's role in 107–108; post-submittal review actions 130–131; quality assurance during design 123–126; resolving differences of opinions 126–128; scope creep 128–130. *see also* quality management

quality assurance model 98–106

quality management 6–7, 97–146; activity scheduling 51–52; overview 97–98; project quality assurance model 101–102; project quality assurance model application 104–106; quality assurance model 98–106; quality definitions 99–100; quality management spectrum 102–104. *see also* quality, design

quality management, construction. *see* construction quality

quality objectives 108–118; aesthetic 116–118; cost 108–113; functional 115–116; performance 113–115

retainage 230–235; design-builder's perspective 231–232; options for 232–234

safety oversight 18

scheduling 37–63; activity definition and coordination 44–62; analyzing contract requirements 39–42; baseline schedule 59–60; builder activity 52–53; conceptual schedule validation 43; conference 42–62; construction integration activities 47–48; contract clauses 40–41; coordinating with the contract 55–56; cost and resource loading 57–59; cost engineering cross-check activities 48–51; designer-of-record activity 46–52; hierarchy of schedules 59–61; imperatives 62–63; implementing 42; liquidated damages and incentives 56–57; look-ahead schedule 60–61; overview 37–39; owner activity 44–46; partnered schedule integration 53–59; quality management activities 51–52; subcontractor input 59; updates 61–62

scope creep 128–130, 174–175

sustainability 273–279; commissioning 277–278; construction waste management 274–275; indoor air quality 275–276; local adaptation 274; material selection 274–275; regional adaptation 274; requirements applied after contract award 279; site work 276–277; voluntarily incorporated 278–279

transportation projects 2–3, 4–5; detractors of design-build 6–8

trust 18–20

value analysis 80–81

value engineering 80–81

About the Authors

James E. Koch, Ph.D., P.E., is the Chief Executive Officer of Quasco LLC, a specialty contractor and professional services firm with offices on multiple continents. He received a B.S. in commerce from Washington and Lee University, an M.S. in civil engineering from Stanford University, the degree of Engineer in Construction Engineering and Management from Stanford University, and a Ph.D. from the University of Missouri-Rolla in Engineering Management. He is a registered Professional Engineer in Civil Engineering in the District of Columbia and is a LEED 2.0 accredited professional. Dr. Koch was an engineer officer in the United States Army serving world-wide in a broad range of staff and command assignments; he retired as a lieutenant colonel after a 21-year career. Since his retirement, Dr. Koch has worked in multiple fields for both contractors and owners as a program and project manager. His leadership positions have ranged from serving as the project manager for construction of the St. Louis Metro's St. Clair County light rail project to serving as a senior vice president for Hill International, where he was program manager for the joint venture responsible for construction management support to the Iraq Reconstruction Programs. In 2005, he was named the Hill International Project Manager of the Year. He served as the director of a university technology transfer center focused on the use of new construction materials and techniques, served as the managing director of the Concrete Industry's Strategic Development Council, and served as the director of the Construction Management Program at Washington University in St. Louis. Dr. Koch's primary focus remains in the areas of program controls, construction in contingency operations, and construction quality.

Douglas D. Gransberg, Ph.D., P.E., is the Donald F. and Sharon A. Greenwood Professor of Construction Engineering at Iowa State University. He received both his B.S. and M.S. degrees in civil engineering from Oregon State University and his Ph.D. in civil engineering from the University of Colorado at Boulder. He is a registered Professional Engineer in Oklahoma, Texas, and Oregon; a Certified Cost Engineer; a Designated Design-Build Professional; and a chartered Quantity Surveyor of the Royal Institution of Chartered Surveyors in the United Kingdom. Before joining academe in 1994, he spent more than 20 years as an officer in the U.S. Army Corps of Engineers, retiring at the rank of Lieutenant Colonel. In his final posting, Dr. Gransberg was the Europe District's Area Engineer stationed in

Ankara, Turkey, where he pioneered the use of design-build and design-bid-build best-value contracting to deliver facilities in remote locations. In 1997, he completed his first design-build research contract with the Texas Department of Transportation to develop the documentation necessary to implement design-build upon legislative approval. In the past 16 years, his research has encompassed the full breadth of alternative project delivery and its applications in the United States and internationally. He has been on the Oklahoma Governor's Task Force on Construction Law since 1999 and helped develop legislation authorizing innovative project delivery practices for public building projects in that state. Dr. Gransberg has been recognized with the AACE International Teaching Excellence Award (2006), the Society of American Military Engineers Toulmin Medal (2003), and the Transit New Zealand Road Innovation Research Award (2005), as well as having been named an Erskine Fellow and a Fulton Hogan Research Fellow at the University of Canterbury in New Zealand (2009).

Keith R. Molenaar, Ph.D., is the department chair and K. Stanton Lewis Chair of Construction Engineering and Management in the Department of Civil, Environmental, and Architectural Engineering at the University of Colorado at Boulder. His research focuses on alternative project delivery strategies for constructed facilities and infrastructure. He maintains an active research program with funding from a variety of public agencies and private companies in both the building and heavy infrastructure sectors. Dr. Molenaar holds a B.S. in architectural engineering and M.S. and Ph.D. degrees, all from the University of Colorado. He was previously a faculty member at the Georgia Institute of Technology, where he was group leader of the Construction Research Center's Procurement and Project Delivery research initiative. Upon completion of his B.S., he worked for Architectural Resource Consultants, Inc., where he specialized in preconstruction planning for owners and designers. Dr. Molenaar was recognized as a "Top 50 Trendsetter" by *Public Works* magazine in 2004; he received the Society of American Military Engineers Toulmin Medal (2003), the American Society of Civil Engineers Thomas Fitch Rowland Award (2001), and the Design-Build Institute of America Academic Leadership Award (2000).